NASA STI Program . . . in Profile

Since its founding, NASA has been dedicated to the advancement of aeronautics and space science. The NASA scientific and technical information (STI) program plays a key part in helping NASA maintain this important role.

The NASA STI program operates under the auspices of the Agency Chief Information Officer. It collects, organizes, provides for archiving, and disseminates NASA's STI. The NASA STI program provides access to the NASA Aeronautics and Space Database and its public interface, the NASA Technical Report Server, thus providing one of the largest collections of aeronautical and space science STI in the world. Results are published in both non-NASA channels and by NASA in the NASA STI Report Series, which includes the following report types:

- TECHNICAL PUBLICATION. Reports of completed research or a major significant phase of research that present the results of NASA programs and include extensive data or theoretical analysis. Includes compilations of significant scientific and technical data and information deemed to be of continuing reference value. NASA counterpart of peer-reviewed formal professional papers, but having less stringent limitations on manuscript length and extent of graphic presentations.

- TECHNICAL MEMORANDUM. Scientific and technical findings that are preliminary or of specialized interest, e.g., quick release reports, working papers, and bibliographies that contain minimal annotation. Does not contain extensive analysis.

- CONTRACTOR REPORT. Scientific and technical findings by NASA-sponsored contractors and grantees.

- CONFERENCE PUBLICATION. Collected papers from scientific and technical conferences, symposia, seminars, or other meetings sponsored or co-sponsored by NASA.

- SPECIAL PUBLICATION. Scientific, technical, or historical information from NASA programs, projects, and missions, often concerned with subjects having substantial public interest.

- TECHNICAL TRANSLATION. English-language translations of foreign scientific and technical material pertinent to NASA's mission.

Specialized services also include creating custom thesauri, building customized databases, and organizing and publishing research results.

For more information about the NASA STI program, see the following:

- Access the NASA STI program home page at *http://www.sti.nasa.gov*

- E-mail your question via the Internet to help@sti.nasa.gov

- Fax your question to the NASA STI Help Desk at 443-757-5803

- Phone the NASA STI Help Desk at 443-757-5802

- Write to:
  NASA STI Help Desk
  NASA Center for AeroSpace Information
  7115 Standard Drive
  Hanover, MD 21076-1320

NASA/TM-2009-215732
NESC-RP-05-67/04-069-I

Cassini/Huygens Probe Entry, Descent, and Landing (EDL) at Titan Independent Technical Assessment

Richard W. Powell, Mary Kae Lockwood, Juan R. Cruz,
Scott A. Striepe, Brian R. Hollis and Michael J. Wright
Langley Research Center, Hampton, Virginia

Jere Justus and Aleta Duvall
Morgan Research Corporation, Huntsville, Alabama

Vernon W. Keller
Marshal Space Flight Center, Huntsville, Alabama

Deepak Bose and Dinesh Prabhu
ELORET Corporation, Ames Research Center, Moffett Field, California

Y. K. Chen and Joe Olejniczak
Ames Research Center, Moffett Field, California

Kenneth Sutton and Jody Fisher
National Institute of Aeronautics, Langley Research Center, Hampton, Virginia

Naruhisa T. Takashima
Analytical Mechanics Association, Langley Research Center, Hampton, Virginia

National Aeronautics and
Space Administration

Langley Research Center
Hampton, Virginia 23681-2199

May 2009

Available from:

NASA Center for AeroSpace Information
7115 Standard Drive
Hanover, MD 21076-1320
443-757-5802

Cassini/Huygens Probe Entry, Descent, and Landing (EDL) at Titan

Independent Technical Assessment

Performed and Prepared by

The NASA Engineering and Safety Center (NESC)

May 26, 2005

NESC Request No. 04-069-I

ACKNOWLEDGEMENTS

In addition to the NESC ITA Assessment team, the team would like to recognize the following individuals from the JPL Huygens Mission Risk Reduction EDL (MRR) team. The NESC and the MMR teams worked together to assess the overall Huygens EDL risks.

JPL Huygens Mission Risk Reduction EDL Team

Gentry Lee, Lead
Wayne Lee
Rob Manning
Christine Szalai

NASA Engineering and Safety Center Technical Assessment Report	Document #: RP-05-67	Version: 1.0
Title: **Independent Technical Assessment of Cassini/Huygens Probe Entry, Descent and Landing (EDL) at Titan**		Page #: 3 of 116

TABLE OF CONTENTS

VOLUME I: ITA REPORT

ACKNOWLEDGEMENTS 2

1.0 AUTHORIZATION AND NOTIFICATION 7

2.0 SIGNATURE PAGE (ASSESSMENT TEAM MEMBERS) 8

3.0 TEAM MEMBERS, EX OFFICIO MEMBERS, AND CONSULTANTS 9

4.0 EXECUTIVE SUMMARY 10

5.0 OVERVIEW OF ITA PLAN 13
    5.1 Schedule, Milestones, Information Requirements, Products Produced 15

6.0 DESCRIPTION OF THE PROBLEM, PROPOSED SOLUTIONS, AND RISK ASSESSMENT 17
    6.1 Problem 17
    6.2 Proposed Solutions 17
    6.3 Risk Assessments 17

7.0 DATA ANALYSIS 20

8.0 OBSERVATIONS AND RECOMMENDATIONS 22

9.0 DEFINITION OF TERMS AND ACRONYMS 23

10.0 LESSONS LEARNED 25
    10.1 Mission Design 25
    10.2 Aerothermodynamics 25
    10.3 Analysis Process 26

VOLUME II: APPENDICES

Appendix A ITA/I Request Form (NESC-FM-03-002) 30

Appendix B Huygens Aerodynamic Database 32
B.0 Introduction 32
B.1 Aerodynamic Database Modifications 33
B.2 Uncertainties 36
B.3 Summary 38

Appendix B Figures
B-1 Huygens and Genesis Sample Return Capsule Configurations 33
B-2 Free Molecular/Modified Newtonian Analyses 34
B-3 Nominal CA variation with Mach Number 35
B-4 Pitch Damping Coefficients 36
B-5 Nominal CA vs. Mach Number with ESA Uncertainty 37
B-6 Pitch Damping Coefficients with Uncertainties 38

NASA Engineering and Safety Center
Technical Assessment Report

Document #:
RP-05-67

Version:
1.0

Title:

Independent Technical Assessment of Cassini/Huygens Probe Entry, Descent and Landing (EDL) at Titan

Page #:
4 of 116

Appendix B Tables
B-1   Aerodynamic Uncertainties 37

Appendix C   Huygens Parachute System Evaluation 39
C.0   Introduction.. 39
C.1   Parachutes Drag Models 40
C.2   Parachute Opening Loads and Structural Strength 42
C.3   Descent Time Calculations and Results.. 43

Appendix C Figures
C-1   Huygens EDL Sequence 40

Appendix C Tables
C-1   Key Parameters for the Huygens Parachutes 41
C-2   Huygens Parachute Drag Model Sets.. 42
C-3   Pilot and Main Parachute Opening Loads and System Strength Summary 43
C-4   Summary of Final Descent Time Calculations 44

Appendix D   Atmospheric Properties.. 46
D.0   Introduction.. 46

Appendix D Figures
D-1   Minimum/Average/Maximum Density Envelope.. 46
D-2   Examples of Output from Titan-GRAM's "High Frequency" Perturbation Model.................... 47
D-3   Comparison of Density Profiles from Hourdin et al. (1995) GCM and ISO Profile
        (Coustenis et al, 2003) with original Yelle Density Profiles 48
D-4   Comparison of Nominal (0% Argon, 1.9% methane mole fraction) TAMWG T0 Density Profile
        (Yelle, 2004) with original Yelle et al. (1997) Density Profiles... 49
D-5   Comparison of TAMWG T0 Density Profiles 0 through 5 (Yelle, 2004a) with Original
        Yelle et al. (1997) Density Profiles... 50
D-6   Comparison of TAMWG T0 (nominal) and TA (maximum methane) Temperature Profiles....... 51
D-7   Comparison of TAMWG T0 (nominal) and TA (maximum methane) Density Profiles
        (Yelle, 2004a) with original Yelle et al. (1997) Density Profiles 52
D-8   Profiles of Equatorial Mean Eastward Wind and Wind Perturbation Standard Deviations,
        from the Titan-GRAM Model (Justus et al., 2004)... 53

Appendix D Tables
D-1   TAMWG October 6, 2004 Titan T0 Profiles and Mole Fractions of Molecular Nitrogen (N_2),
        Argon (Ar), and Methane (CH_4), in Percent 50

Appendix E   POST2-Based Flight Simulation... 62
E.0   Introduction.. 62
E.1   Core Trajectory Simulation in POST2.. 63
E.1.1   Flight and Systems Studies Heritage of POST2 63
E.1.2   6DOF Entry-3DOF Parachute Descent Trajectory 63

NASA Engineering and Safety Center
Technical Assessment Report

Document #:
RP-05-67

Version:
1.0

Title:
Independent Technical Assessment of Cassini/Huygens Probe Entry, Descent and Landing (EDL) at Titan

Page #:
5 of 116

E.2   Initial States for Entry Vehicle 65
E.3   Entry Vehicle Aerodynamics Database Subroutine 66
E.4   Parachute Model Characteristics 67
E.5   Titan-GRAM Atmospheric Model 67
E.6   Huygens Probe Mass Property Data 72
E.7   Aeroheating Parameter Calculations 72

Appendix E Figures
E-1   Nominal 6DOF-3DOF Descent Profile 64
E-2   Nominal 6DOF-3DOF Entry Phase Characteristics 65
E-3   GCM Perturbed Density % from Nominal 69
E-4   TA Profile Perturbed Density % from Nominal 70
E-5   TA Profile Perturbed EW Winds from Nominal 71
E-6   TA Profile Perturbed NS Winds from Nominal 72

Appendix E Tables
E-1   POST2 Simulation Inputs 66
E-2   Titan-GRAM Inputs 68

Appendix F   Monte Carlo Analyses and Results 73
F.0   Introduction 73
F.1   Monte Carlo Analysis 73
F.2   Monte Carlo Dispersed Inputs 73
F.3   Monte Carlo Results 75
F.3.1   Parachute Opening Loads 76
F.3.2   Total Time-Of-Flight 77
F.3.3   Aerothermal Indicator Bounding Cases 79
F.3.4   High Angle-of-Attack at Peak Heating Case 80
F.3.5   Entry Sensitivity Assessment 81

Appendix F Figures
F-1   Main Parachute Peak Opening Load using ESA Parachute Drag Model 76
F-2   Main Parachute Peak Opening Load using NASA 1 Parachute Drag Model 77
F-3   Time of Impact from Pilot Parachute Deployment using Vorticity 1 Parachute Drag Model 78
F-4   Time of Impact from Pilot Parachute Deployment using NASA 2 Parachute Drag Model 79
F-5   Maximum Heat Rate from Monte Carlo Output 80
F-6   Maximum Total Angle-of-Attack at Maximum Heat Rate 81

Appendix F Tables
F-1   Huygens Titan Probe 6DOF Entry Dispersions 74
F-2   Monte Carlo Cases Completed 75

Appendix G   Aerothermodynamics 82
G.0   Introduction 82
G.1   Background 82

NASA Engineering and Safety Center
Technical Assessment Report

Document #: RP-05-67

Version: 1.0

Title:

Independent Technical Assessment of Cassini/Huygens Probe Entry, Descent and Landing (EDL) at Titan

Page #: 6 of 116

G.2   Methodology 83
G.3   Key Results 93
G.4   Conclusions 94

Appendix G Figures
G-1   11-Nov-2004 Max Heat-Rate and Max Heat-Load Trajectories from POST 85
G-2   Nominal Convective Heating Time-History at Stagnation Point on Rate Trajectory 86
G-3   Nominal Radiative Heating Time-History at Stagnation Point on Rate Trajectory 87
G-4   Nominal Total Heating Time-History at Stagnation Point on Rate Trajectory 87
G-5   Heat-Rate at Stagnation Point 88
G-6   Heat-Load at Stagnation Point 89
G-7   Heat-Rate at Mid-Cone 90
G-8   Heat-Load at Mid-Cone 91
G-9   Example of Diffusion-Modeling Effects on Convective Heating 92

Appendix H   Thermal Protection System 98
H.0   Introduction 98
H.1   Background 98
H.2   UV Lamp 100
H.3   UV Test Procedure 102
H.4   UV Tests 102
H.4.1   UV Test at 50 W/cm^2 102
H.4.2   Convective Cooling Effects 104
H.4.3   UV Test at 150 W/cm2 105
H.5   Arc Jet Test 107
H.6   Analytical Modeling 109
H.6.1   Recession Modeling 109
H.6.2   Thermal Modeling 114

Appendix H Figures
H-1   AQ60 Sample Geometry for UV Radiation Tests 99
H-2   Predicted Radiation for the Titan Probe and the Spectral Distribution for the Test Facilities
        Considered 101
H-3   Thermal Data from UV Test at 50 W/cm^2 103
H-4   Photographs of AQ60 Before and After at Exposure at 50 W/cm^2 for 150 Seconds 104
H-5   Effect of N$_2$ Gas Flow across AQ60 char at 50 W/cm^2 105
H-6   Test Data Showing In-Depth Thermocouple Response at 150 W/cm^2 106
H-7   Photos of AQ60 Before and After Test at 150 W/cm^2 for 30 Seconds 107
H-8   Pyrometer and Thermocouple Data from Arc Jet Test of AQ60 108
H-9   Photos of AQ60 Before and After Arc Jet Exposure at 80 W/cm^2 108
H-10   Empirical Correlation of AQ60 Arc Jet Recession Data 110
H-11   Comparison of Thermochemical Ablation Model with IRS Arc Jet Data in Simulated Titan
        Atmosphere 112
H-12   Comparison of Thermochemical Ablation Model with IRS Arc Jet Data in N$_2$ 113

	NASA Engineering and Safety Center Technical Assessment Report	Document #: RP-05-67	Version: 1.0
	Independent Technical Assessment of Cassini/Huygens Probe Entry, Descent and Landing (EDL) at Titan	Page #: 7 of 116	

H-13   Comparison of Non-Dimensional Mass Loss Rates at Two Pressures 114

1.0   AUTHORIZATION AND NOTIFICATION

Mr. David Leckrone, NASA Engineering and Safety Center (NESC) Chief Scientist, initiated the request for the NESC to conduct an independent technical assessment (ITA) regarding the Huygens entry, descent and landing (EDL) into Titan. This request was presented to the NESC Review Board (NRB) on August 12, 2004.   Mr. Matt Landano, NESC JPL Chief Engineer, concurred outside the board that this is an appropriate and necessary NESC activity. The NRB approved the initiation of an ITA at the same meeting.

The ITA Plan was developed by Mr. Richard Powell and approved by the NRB on August 26, 2004.

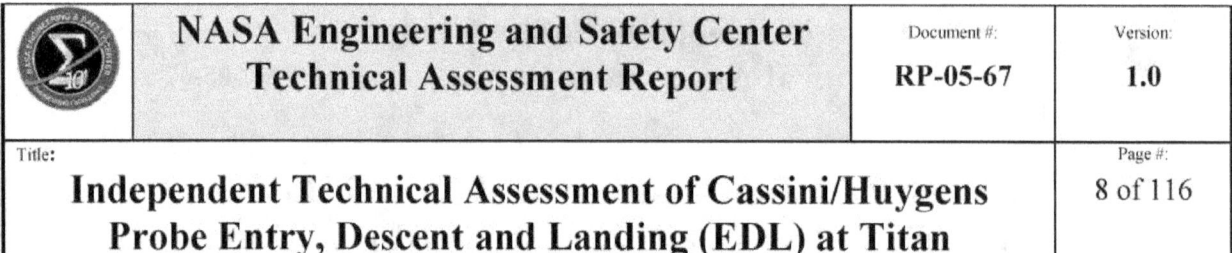
2.0   SIGNATURE PAGE (ASSESSMENT TEAM MEMBERS)

Richard W. Powell, Co-Lead	Mary Kae Lockwood, Co-Lead
C. G. (Jere) Justus	Vernon Keller
Deepak Bose	Dinesh Prabhu
YK Chen	Joe Oleiniczak
Juan R. Cruz	Scott A. Striepe
Aleta Duvall	Kenneth Sutton
Jody Fisher	Naruhisa Takashima
Brian Hollis	Mike Wright
Bernie Laub	

3.0   TEAM MEMBERS, EX OFFICIO MEMBERS, AND CONSULTANTS

Team Members	**Discipline**
Richard W. Powell, LaRC	Co-Lead, Flight Mechanics and Trajectory Simulation
Mary Kae Lockwood, LaRC	Co-Lead Systems and Aerodynamics/ Aerothermodynamics
Brian Hollis, LaRC	Aerothermodynamics
Juan R. Cruz, LaRC	Parachutes
Scott A. Striepe, LaRC	Flight Mechanics and Trajectory Simulation
Naruhisa Takashima, LaRC/AMA	Aerodynamics/Aerothermodynamics
Kenneth Sutton, LaRC/NIA	Aerothermodynamics
Jody Fisher, LaRC/NIA	Flight Mechanics and Trajectory Simulation
Aleta Duvall, MSFC/Morgan Research	Atmospheric Modeling
C. G. (Jere) Justus, MSFC/Morgan Research	Atmospheric Modeling
Vernon Keller, MSFC	Atmospheric Modeling
Mike Wright, ARC	Aerothermodynamics
Deepak Bose, ARC/ELORET	Aerothermodynamics
Y-K Chen, ARC	TPS Response Model
Dinesh Prabhu, ARC/ELORET	Aerothermodynamics/Radiative Heating
Joe Olejniczak, ARC	Aerothermodynamics
Bernie Laub, ARC	Thermal Protection System

Consultants	
Steven Labbe, NESC/JSC	NDE/Flight Sciences
Claude Graves, JSC	Flight Mechanics/Flight Systems
Chris Cerimele, JSC	Flight Mechanics/Flight Systems
Pete Cuthbert, JSC	Parachute Performance
Dean Kontinos, ARC	Aerothermodynamics/Flight Systems

NASA Engineering and Safety Center Technical Assessment Report	Document #: RP-05-67	Version: 1.0
Title: **Independent Technical Assessment of Cassini/Huygens Probe Entry, Descent and Landing (EDL) at Titan**		Page #: 10 of 116

4.0 EXECUTIVE SUMMARY

Starting in January 2004, the NESC has received several communications from knowledgeable technical experts at NASA expressing shared concerns (mainly at the Langley Research Center (LaRC) and Ames Research Center (ARC)) about Huygens mission success. It was suggested that NASA become more technically involved directly in the analysis of Huygens' entry, descent and landing (EDL) focusing on the following primary concerns:

1. The parachute deployment trigger performance and the resultant effects on the operation of the parachute system, and

2. The determination of the radiative heating environment at Titan by ESA and the corresponding thermal protection system (TPS) response.

A NESC Team was formed and tasked to provide an independent assessment of these concerns. Tasks required for this independent assessment included:

1. Development of an aerodynamics database for the entry configuration.
2. Development of an aerodynamic database for the probe under each of the three parachutes.
3. Development of an opening loads model for each of the three parachutes.
4. Development of detailed aerothermal environments. These environments were comprised of the convective (laminar and turbulent) and radiative heat rates and integrated heat loads.
5. Update of the Titan atmosphere model to include the Cassini observation of Titan on November 15, 2004 (Project has denoted this as the "TA" pass).
6. Development of a Monte Carlo trajectory simulation analysis capability.
7. Determination of entry flight path angle that maximizes the mission success probability.
8. Determination of the TPS evaluation trajectories.
9. Determination of applicability and appropriateness of radiative heating models.
10. TPS performance assessment.

The NESC team used tools and atmospheric observation data to evaluate the Huygens EDL. Primary issues and findings identified from this assessment were:

1. The NASA Monte Carlo peak opening loads for the pilot and main parachutes were below the stated strength of these parachutes at the 99.87 percentile level. Thus the risk to the mission due to structural failure associated with the pilot or main parachute opening was retired as a concern.

2. Using the final NASA entry vehicle and parachute drag models and the atmospheric profiles generated from the TA encounter, the design requirement time-of-flight was exceeded by approximately 24% of the Monte Carlo simulations. Analyses conducted by ESA (Vorticity) showed that the time-of-flight requirement was exceeded in 0.4% to 4% of the Monte Carlo simulations. The ESA was aware of this difference and elected to accept the potential for flight times longer than the design requirement rather than take any corrective action.

3. Through comparison of the NESC aerothermodynamic analysis, with that performed by ESA to support the final probe release decision, it was determined that the ESA analysis included two significant potentially non-conservative assumptions which were not made in the NESC analysis. First that a non-equilibrium distribution of excited states of the CN molecule would result in lower radiative heating rates than would be predicted by an equilibrium model. Second, that boundary layer transition, with the accompanying turbulent heating augmentation, would not occur until well after peak heating on the entry trajectory and thus would not significantly affect the heating environment.

4. The uncertainties applied in the ESA's aerothermodynamic analysis were less than those which normally would be applied by NASA during the design process and thus presented a less conservative estimate of mission risk. Fortunately, measurements of the upper level of CH_4 concentration in Titan's atmosphere, which determines the amount of CN radiation produced in the vehicle's bow-shock wave, during the Cassini T0 and TA passes, were lower than previous estimates (reduced from 5.0% to 2.3%). Thus, the actual aerothermodynamic environment was less severe than previously thought. Although fortuitous, this finding was a post-design reduction in risk rather than a designed conservatism.

5. The NESC analysis was more conservative than that of the ESA in regards to radiation and transition modeling and overall uncertainties (although still less conservative than current best practice at NASA). The conclusion of this analysis was that the potential existed for the integrated heat load to exceed the design specification of 4000 J/cm^2.

6. The ESA analysis assumed that the Huygens probe angle-of-attack at peak heating would be zero. The NASA Monte Carlo results showed that an angle-of-attack as high as 5 degrees (3-sigma) is possible. Using the trajectory from this 3-sigma case, the aerothermodynamic environment analysis indicated that this high angle-of-attack at peak heating would produce some heating amplification when compared to the ESA assumed attitude at peak heating.

NASA Engineering and Safety Center Technical Assessment Report	Document #: RP-05-67	Version: 1.0
Title: **Independent Technical Assessment of Cassini/Huygens** **Probe Entry, Descent and Landing (EDL) at Titan**		Page #: 12 of 116

However, the team concluded that the amount of heating amplification would not raise the risk above the low-to-moderate range.

7. Preliminary shock tube data (Bose et al., AIAA Paper 2005-0768) on the radiation produced by the CN molecule in the hypersonic flow regime indicated the possibility of over-prediction in the radiative heating rates for the Huygens entry using the current models and assumptions. These tests were conducted by NASA ARC in 2004 independently of the NESC activity, and the preliminary data were analyzed and released only days before the probe release decision. The team elected not to reduce the risk rating since the detailed assessment of the data had not been completed.

As part of the tasks, various mission parameter sensitivity assessments were completed for the Huygens probe entry to determine the potential for increasing margins in maximum heating rate, heat load, angle-of-attack at atmospheric interface, time-of-flight, etc. These assessments included:

- changing the mean flight path angle at atmospheric interface,
- reducing the flight path angle uncertainty at atmospheric interface, and
- varying the nominal probe target orientation at atmospheric interface.

All these assessments showed only small sensitivities for the mission margins to the assessment parameters. These small sensitivities were traced to the large atmospheric density scale height of Titan (~40 km as compared to ~7 km for Mars and Earth). Therefore, the final recommendation was to not change any mission parameter. Because of this assessment, the assessments led by the Jet Propulsion Laboratory (JPL), and the assessments led by the ESA, this recommendation was adopted and the Huygens probe was released nominally on December 24, 2004.

	NASA Engineering and Safety Center Technical Assessment Report	Document #: RP-05-67	Version: 1.0
	Title: Independent Technical Assessment of Cassini/Huygens Probe Entry, Descent and Landing (EDL) at Titan		Page #: 13 of 116

5.0 OVERVIEW OF ITA PLAN

The Cassini/Huygens mission is a joint program of NASA and ESA and is managed by NASA JPL. The Huygens probe portion of the mission is ESA's responsibility. The Cassini/Huygens spacecraft is currently in orbit around Saturn, following the successful Saturn Orbital Insertion (SOI) burn on July 1, 2004 (UTC)[1]. As scheduled, the Huygens probe separated from the Cassini spacecraft on December 25, 2004 (UTC). On January 14, 2005 (UTC), it entered the atmosphere of the moon Titan, descended on parachutes, and landed on Titan's surface. It was in NASA's interest to provide whatever assistance is possible to help assure the success of Huygens.

The scope of the ITA, at the direction of NESC, was to conduct an independent assessment of the EDL of the Huygens probe. The assessment was comprised of a combination of review of relevant data and independent analyses to perform an independent EDL assessment of the Huygens probe entry into Titan. The assessment involved the development of a full EDL simulation model for Huygens, as though it were NASA's responsibility. To the extent that the International Traffic at Arms Regulation (ITAR) made it difficult to provide NASA flight data to the ESA, these data could only be fully incorporated in the models developed by NASA. Note that the analyses do require data input that only the ESA could provide.

The ITA team participated in assessing the residual risk to the Huygens probe prior to the next go/no-go point in the schedule. The schedule included finalizing any modifications to Huygens probe EDL on-board software in the mid-November 2004 timeframe. Because this effort was initiated so near to the probe release date, the analyses focused on specific areas of concern and should not be taken to represent a comprehensive, system-wide review. The areas of focus were: aerodynamics, parachute performance analysis, atmospheric properties, flight mechanics, aerothermodynamics, and TPS. A description of each discipline and the work involved is detailed below.

Aerodynamics

A high-fidelity, 6 degree-of-freedom (DOF) aerodynamics database was developed to support simulations from the Titan atmospheric interface to touchdown of the Huygens probe. This aerodynamics database was developed using data from previous NASA flight vehicles similar to the Huygens shape (e.g. Genesis, MER, Phoenix), data from ballistic range testing of the Huygens probe configurations, and detailed computational fluid dynamics (CFD) models of the Huygens probe at selected atmospheric flight conditions. Based on these calculations and

---

[1] Universal Coordinated Time equivalent to Greenwich Mean Time

NASA Engineering and Safety Center
Technical Assessment Report

Document #:	Version:
RP-05-67	1.0

Title:

Independent Technical Assessment of Cassini/Huygens Probe Entry, Descent and Landing (EDL) at Titan

Page #:

14 of 116

engineering assessments, aerodynamic uncertainties were also determined to support Monte Carlo simulation analyses.

Parachute Drag Models

Four sets of parachute drag models were provided to the Flight Mechanics team. The first set was supplied by ESA (ESA Model). The second set was supplied by Vorticity (Vorticity 1 Model). The third set was generated by NASA using data on similar parachute configurations on NASA probes (NASA 1 Model). The fourth set was developed by NASA using wind tunnel and flight test data from the Huygens parachute system development program (NASA 2 Model). These models (including uncertainties) were used to estimate opening loads and time-of-flight. The NASA models were created using an approach similar to those being implemented by the Mars Phoenix and the Mars Science Laboratory (MSL) programs. In general, the NASA parachute drag models had higher mean drag coefficient values and larger uncertainties than those used by ESA and its contractors.

Atmosphere Properties

A NESC team member attended Titan-0 (T0) and Titan-A (TA) flyby workshops sponsored by the Titan Atmospheric Model Working Group (TAMWG). The TAMWG expeditiously provided atmospheric data results based on conclusions and recommendations of these workshops. Updates based on these results were included in the atmospheric model, in the aerothermodynamics environment models, and the flight simulation. The primary impact of the T0 and TA data was a reduction in the estimated concentration of CH_4 in the atmosphere, which lowered the predicted radiative heating component.

Flight Mechanics

A trajectory simulation composed of a 6DOF entry followed by a 3DOF parachute phase was developed. This simulation included the aerodynamics database, the atmospheric models, and the parachute performance models (both ESA and NASA) described above. The aerothermo-dynamics environments team also provided heating environment models (estimates of laminar, turbulent, and radiative aeroheating) that were included in the simulation. JPL provided the nominal initial states and the associated covariance so that a Monte Carlo evaluation could be performed. The simulation outputs included statistics on peak heat rate, heat load, angle-of-attack at peak heating and parachute deploy, parachute opening loads, landing location, and total time-of-flight for the probe.

**NASA Engineering and Safety Center
Technical Assessment Report**

Document #:
RP-05-67

Version:
1.0

Title:
**Independent Technical Assessment of Cassini/Huygens
Probe Entry, Descent and Landing (EDL) at Titan**

Page #:
15 of 116

Aerothermodynamics

The aerothermodynamic environment during entry was predicted using computational methods (flow field and radiation transport codes) for the forebody of the vehicle. An agreement on aerothermodynamic methodology was reached between the NESC and ESA teams during the November 4-5, 2004 Aerothermal Convergence Working Group meeting at ESA headquarters in Paris, France. Because of time constraints and the difficulty of performing wake flow simulations, only a minimal analysis of the aftbody environment and off-nominal angle-of-attack conditions were performed. The result of these computations was time-histories of the heating rate and integrated heat loads along trajectories identified by the 6DOF simulations as being the worst-cases for maximum heat rate and maximum heat load. These time histories were supplied to ESA to be used as inputs for thermal response modeling of the vehicle's heat shield. Details of the aerothermodynamic analysis are presented in Appendix G.

Thermal Protection System

The performance of the Huygens forebody TPS material, AQ60[2], when exposed to ultraviolet (UV) radiation, was evaluated by testing well-instrumented samples with a UV lamp. This work was sponsored and supported by NASA's In-Space Propulsion (ISP) Aerocapture project. Under this ITA, work was extended by conducting an arc jet test of a well-instrumented sample of AQ60 and comparing the temperature data with predictions using the European Aeronautic Defence and Space Company (EADS) model for this material. In addition, European arc jet data acquired during Huygens development were modeled with thermochemical ablation theory to evaluate the risk of extrapolating an empirical fit of these data to the flight environment where pressures would be approximately four times greater than ground test.

5.1 Schedule, Milestones, Information Requirements, Products Produced

Objective: To perform an independent EDL assessment of the Huygens probe entry into Titan.

Schedule: August 26, 2004 – December 22, 2004

Milestones:

August 12, 2004   NRB Approval of ITA/I
August 12-25, 2004   ITA Team Formation
August 26, 2004   NESC ITA Approved/Initiated

---

[2] A proprietary product of the EADS

August 27, 2004	ITA Team Kickoff Meeting (Telecom)
September 8-9, 2004	Titan Atmosphere Model T0 Workshop
November 4-5, 2004	ESA/NASA Aerothermodynamics Workshop
November 9-11, 2004	ESA/NASA EDL Workshop
November 11, 2004	Initial EDL Risk Assessment Completed
November 15, 2004	Titan Atmospheric Workshop – TA Results
November 29, 2004	Presentation of EDL Risk Assessment to ESA/NASA (Joint with JPL Mission Risk Review Team)
December 24, 2004	Nominal Huygens Probe Separation from Cassini Spacecraft
January 14, 2005	Nominal Huygens Probe Entry into Titan
February 28, 2005	Draft of Final Evaluation Report Provided To Peer Review Panel
May 26, 2005	Final Evaluation Report presented to the NRB

Information Required:

1.   Obtain the following data/inputs from ESA:

   a.   Entry vehicle outer mold line geometry.
   b.   Mass properties.
   c.   Parachute description.
   d.   Description of parachute deployment triggers.
   e.   Aerodynamic database.
   f.   Aeroshell design conditions.
   g.   Aeroshell Aerodynamic heating assumptions.
   h.   TPS material (AQ60) properties data.
   i.   Nominal initial conditions at Titan entry interface.

2.   Products

   a.   Development of aerodynamic database.
   b.   Comparison of ESA/NASA parachute models.
   c.   Parachute opening loads statistics.
   d.   Modifications (if any) to Titan Global Reference Atmospheric Model (Titan-GRAM), atmospheric properties model, developed by NASA to support Titan aerocapture systems studies) to support lander mission.
   e.   Updates to Titan-GRAM after Cassini's TA pass.
   f.   Touchdown location and time-of-flight statistics from pilot parachute deployment to landing.
   g.   Sensitivity of entry flight path angle and dispersion range on vehicle performance.
   h.   Determination of maximum expected heating trajectories from Monte Carlo analysis.
   i.   Comparison of these maximum expected heating trajectories with heating design point provided by ESA to NASA Ames.

j. Aerothermodynamic environment assessment.

6.0 DESCRIPTION OF THE PROBLEM, PROPOSED SOLUTIONS, AND RISK ASSESSMENT

6.1 Problem

Starting in January 2004, the NESC has received several communications from knowledgeable technical experts at NASA (both at LaRC and ARC) expressing concerns about Huygens mission success. It was suggested that NASA become more technically involved directly in the analysis of Huygens' EDL focusing on the following primary concerns: 1) the parachute deployment trigger performance and the resultant effects on the operation of the parachute system, and 2) the determination of the radiative heating environment at Titan by the ESA and the corresponding TPS response.

6.2 Proposed Solutions

In August 2002, NASA completed a detailed aerocapture systems analysis at Titan. This multi-center NASA team funded by NASA's ISP Program and led by Dr. Mary Kae Lockwood, developed atmospheric density models, aerodynamic databases, aerothermodynamic environments, TPS design, and high-fidelity flight simulations with Monte Carlo capability. ESA had already contacted the NASA ARC members of this team for aerothermodynamics assistance. In addition, NASA had recently completed the design and development of the parachute for the Mars Exploration Rovers (MER). This parachute system is similar to that used by Huygens. From the experience provided by these activities, the proposed solution was for the NESC to:

1. Form a NASA-wide team using the atmospheric properties, aerothermodynamics and flight mechanics expertise from the Titan aerocapture systems analysis team. Also, add a member who was heavily involved with the MER parachute design and evaluation.

2. Develop independent aerodynamics, aerothermodynamics, and parachute performance and opening loads models, and a flight simulation with Monte Carlo capability.

3. Use the atmospheric properties model that had been provided to ESA by NASA. Note this model was developed by Jere Justus, a member of the original systems analysis team and also a member of the NESC ITA team.

6.3 Risk Assessments

The first area in which a risk assessment was conducted was the aerothermodynamic environment to which the Huygens probe will be exposed. The Huygens probe will be the first entry into the atmosphere of Titan. Titan has a significant amount of CH_4 in the atmosphere

	NASA Engineering and Safety Center Technical Assessment Report	Document #: RP-05-67	Version: 1.0
Title: **Independent Technical Assessment of Cassini/Huygens Probe Entry, Descent and Landing (EDL) at Titan**			Page #: 18 of 116

(original estimates of as high as 5% prior to the Cassini examination, when the upper limit was reduced to 2.3%). The rest of the atmosphere is primarily N_2. When these CH_4 molecules pass through the high-temperature bow-shock wave of the Huygens probe, they will disassociate and recombine, and some CN will be formed. This CN molecule emits radiation at high temperatures which greatly increases the total heating to the spacecraft. As had been demonstrated by the NASA Titan aerocapture systems studies, this augmentation of the convective heating rate, as well as the TPS response to this radiative element, is difficult to determine even with current analytical capabilities, whereas Huygens was designed and launched many years ago. The risk is that the heat shield tools available when Huygens was designed may not have been sufficient to correctly design the Huygens probe to withstand the aerothermodynamic environment. Another aerothermodynamic concern was that the forebody boundary-layer flow would transition to turbulent flow at or before the peak heating point along the trajectory, which would augment the heating to the probe, and could also lead to heat shield failure. Other risks were that the heating to the backshell was underestimated and that the effects of off-nominal (non-zero) angle-of-attack were not assessed.

The second risk area was the parachute. The first risk in the parachute area is that the signal to deploy could be "spoofed" causing the parachute to deploy at the wrong time, or not to deploy at all. The Huygens probe uses sensed acceleration and persistence to trigger the parachute deployment, which is similar to the approach used by the MER. Spoofing was a concern to the MER project, and many Monte Carlo high-fidelity simulations were conducted to quantify the risk of spoofing. The second and third parachute risk areas were parachute opening loads and parachute drag. If the opening loads were underestimated, either the parachute or the probe's support structure could fail. If the drag was underestimated, the time-of-flight would be longer than specified, increasing the risk that the data link with Cassini would be lost before landing.

The path to resolving these issues was to develop an independent set of models and to employ them in simulations to evaluate the potential risks. The models that were developed included an aerodynamic database to support a high-fidelity flight simulation, an aerothermodynamic environment model to support heat-transfer predictions and TPS response evaluation, an atmospheric model, and various parachute models. The aerodynamic database included CFD calculations, data derived from ballistic range tests, and data from other entry capsules. The derivation of this database also included an uncertainty analysis for all parameters that could be used to support a Monte Carlo flight trajectory simulation analysis. In addition to the NASA aerodynamic database, ESA provided its aerodynamic database for comparisons (refer to Appendix B). Aerothermodynamic models were developed by Ames and LaRC, and independent analyses were conducted for convective laminar and turbulent heating and radiative heating of the heat shield (refer to Appendix G). The ARC and LaRC teams compared results from their analyses, and reached consensus on both the mean environment definitions and the uncertainties that should be used. Parachute models for opening loads and performance, along with uncertainties, were developed by NASA. NASA provided ESA with a Titan atmospheric

	NASA Engineering and Safety Center Technical Assessment Report	Document #: RP-05-67	Version: 1.0
Title: **Independent Technical Assessment of Cassini/Huygens Probe Entry, Descent and Landing (EDL) at Titan**		Page #: 19 of 116	

property model that ESA used for its analysis. This same basic model was used for the NASA evaluation with the addition of updates from the two Cassini passes of Titan (refer to Appendix D).

These models as well as estimates of the vehicle position and velocity states at the atmospheric interface (provided by JPL) were employed in the Monte Carlo flight simulations, which were used to screen for stressing cases. These stress cases were then used to provide inputs for further detailed evaluations. The use of Monte Carlo simulations with independent physical models to identify stressing cases is an approach that is used to support NASA's atmospheric flight programs (e.g. MERs) and has been proven to be an effective technique.

7.0   DATA ANALYSIS

Refer to Appendices B through H for additional information.

The atmospheric properties derived from the two Cassini passes (T0 and TA) indicated that the maximum concentration of CH_4 was 2.3%, as compared to the 5% value expected before these observations. The aerothermodynamic models developed by NASA showed that this greatly reduced the level of radiative heating. This reduction, even though partially offset by the expected amplification of turbulent flow and higher uncertainties applied to aerothermodynamic analysis than those by ESA, allowed the NESC team to rate the potential for a TPS failure as low-to-moderate based on the results of ESA TPS thermal response calculations.

When the parachute peak opening loads estimates were compared to the ESA-supplied parachute and structural strength capability, the peak opening load estimates were below the strength capability at the 3-sigma level. The NESC team then rated this risk as acceptable.

The NESC parachute drag models, when incorporated into the Monte Carlo flight simulations, showed that the specified maximum flight time could be exceeded over 24% of the time. Similar simulations by the ESA showed that the flight time was exceeded in less than 4% of the time. ESA understood the NASA assessment and elected to accept the risk to the mission. Because the evaluation of this issue was beyond the ITA charter, the NESC team did not rate this risk.

Many mission parameter modifications were evaluated to determine if the risk to the TPS could be reduced. These modifications included:

1. **Delaying the release of Huygens from Cassini to reduce the entry flight path angle dispersions at Titan atmospheric interface.** Analysis performed by the JPL navigation team showed that delaying the release by 2 days would decrease the uncertainty in the flight path angle from 3 degrees to 1.5 degrees, at the expense of increased fuel for the Cassini deflection maneuver (required to prevent Cassini from entering Titan's atmosphere also).

2. **Modifying the nominal entry flight path angle.** Such a modification could be easily accommodated by Cassini, and ESA had performed mission analysis with entry flight path angles between -62 degrees and -68 degrees. If the analysis had shown a significant advantage for a flight path angle outside this range, the recommendation would have been to delay the release to a later orbit so that all the implications could be studied.

3. **Modifying the deployment trigger algorithm and the subsequent parachute deploy sequence.** Based on NASA's experience with Mars entry vehicles, the possibility that the parachute could deploy at the wrong time must be analyzed since the deployment

NASA Engineering and Safety Center Technical Assessment Report	Document #: RP-05-67	Version: 1.0
Title: **Independent Technical Assessment of Cassini/Huygens Probe Entry, Descent and Landing (EDL) at Titan**		Page #: 21 of 116

trigger has not been designed to handle the full atmospheric density variability and there are uncertainties with aerodynamic characteristics. This phenomenon, namely a combination of atmospheric variability and aerodynamic uncertainties causing the parachute to deploy at the wrong time, is known as spoofing. If modification to the deployment trigger algorithm had shown a significant advantage in reducing the spoofing potential, the recommendation to delay until a future orbit would have been made.

The NESC analyses of the effects of these proposed modifications resulted in the following conclusions:

1. Delaying the release to decrease the flight-path angle uncertainty had minimal effects on the range of aerothermodynamic environments.

2. Within the possible flight-path angle range of -62 degrees to -68 degrees, no change to the nominal conditions could be found to reduce the aerothermodynamic risk.

3. The likelihood of spoofing the parachute release (on the order of 2 cases out of 10,000) was too low to justify modifications to the trigger algorithm to reduce the spoofing potential.

8.0   OBSERVATIONS AND RECOMMENDATIONS

The final recommendation, based on the analysis described in this report, is to not modify the nominal mission design. The NESC team found no options within the design space that would reduce the TPS risk from the low-to-moderate level, and risks associated with the parachute were determined to be too low to warrant any modifications.

9.0   DEFINITION OF TERMS AND ACRONYMS

Acronym	Definition
AIAA	American Institute of Aeronautics and Astronautics
AMA	Analytical Mechanics Associates, Inc.
Ar	Argon
ARC	Ames Research Center
ARD	Atmospheric Reentry Demonstrator
atm	Atmosphere
CFD	Computational Fluid Dynamics
CIRS	Composite Infra-Red Spectrometer
cmq	Pitch Damping Coefficient
DCSS	Descent Control Subsystem
DGB	Disk-Gap-Band
DOF	Degree-Of-Freedom
DMSC	Direct-Simulation Monte Carlo
DPLR	Data-Parallel Line Relaxation – CFD code used at NASA
EADS	European Aeronautic Defence and Space Company
EDL	Entry, Descent And Landing
ESA	European Space Agency
ESTEC	European Space Research and Technology Centre
FAR	Flight Acceptance Review
GCM	General Circulation Model
HQ	NASA Headquarters
INMS	Ion and Neutral Mass Spectrometer
ISO	Infrared Space Observatory
ISS	Imaging Science Subsystem
ISP	In-Space Propulsion
ITA/I	Independent Technical Assessment/Inspection
ITAR	International Traffic in Arms Regulations
JPL	Jet Propulsion Laboratory
JSC	NASA Johnson Space Center
LaRC	NASA Langley Research Center
LAURA	Langley Aerothermodynamic Upwind Relaxation Algorithm - CFD code used at LaRC
MER	Mars Exploration Rovers
MRR	Mission Risk Reduction
MSL	Mars Science Laboratory
MSFC	Marshall Space Flight Center
N_2	Nitrogen

Acronym	**Definition**
NASA	National Aeronautics and Space Administration
NDE	NESC Discipline Expert
NEQAIR	Non-Equilibrium Radiation analysis code used at ARC
NESC	NASA Engineering and Safety Center
NIA	National Institute of Aerospace
NRB	NESC Review Board
POST2	Program to Optimize Simulated Trajectories II
RADEQUIL	Equilibrium Radiation analysis code used at LaRC
RID	Review Item Discrepancy
SOI	Saturn Orbital Insertion
SRC	Sample Return Capsule
T0	JPL's notation for the first non-targeted observation pass of Titan by Cassini. Cassini flyby "0" of Titan (3 July, 2004 at ~ 339000 km).
TA	JPL's notation for the first targeted observation pass of Titan by Cassini. Cassini flyby "A" of Titan (26 October, 2004 at ~ 1200 km).
TB	JPL's notation for second targeted observation pass of Titan by Cassini. Cassini flyby "B" of Titan (13 December, 2004 at ~ 1200 km).
TAMWG	Titan Atmospheric Model Working Group
T/C	Thermocouple
TIM	Technical Interchange Meeting
Titan-GRAM	Titan Global Reference Atmospheric Model
TPS	Thermal Protection System
UTC	Coordinated Universal Time (equivalent to Greenwich Mean Time)
UV	Ultraviolet
UVIS	Ultra-Violet Imaging Spectrograph
VIMS	Visual and Infrared Mapping Spectrometer

	NASA Engineering and Safety Center Technical Assessment Report	Document #: RP-05-67	Version: 1.0
Title: **Independent Technical Assessment of Cassini/Huygens Probe Entry, Descent and Landing (EDL) at Titan**			Page #: 25 of 116

10.0 LESSONS LEARNED

10.1 Mission Design

It was an unexpected insight that modifying the nominal entry flight path angle and reducing variability in the entry flight path angle from ±3 degrees to ±1.5 degrees would have minimal influence on the aerothermodynamic environments. The experience at Earth and Mars has shown a large influence for both the mean and the variability in the entry flight path angle. The root cause of this insensitivity is due to the large atmospheric density scale height of Titan. Atmospheric scale height is the altitude change required to change the density by e. For Earth and Mars, the scale height is on the order of 7-10 km, whereas it is about 40 km at Titan. Because of the large scale height of Titan's atmosphere, the density variation with altitude is small and, thus, the heat load and maximum heat rate are similar over the entire entry flight path angle range that Huygens could encounter.

10.2 Aerothermodynamics

The most challenging aerothermodynamic problem for the Huygens mission was to compute the radiation due to the formation of the radiating molecule CN in the high-temperature bow-shock region. This challenge was related to both the general problem of computing flow fields in which radiative heating levels were high enough to remove significant energy from the flow, (which required a tightly-coupled flow field radiation-transport methodology that is beyond NASA's current capabilities) and to the specific problem of defining the radiative characteristics of the CN molecule in a non-equilibrium environment. Little work has been done in this area since CN radiation has not been a significant concern in any past entry missions such as Apollo or the Mars robotic probes. Because the current methodology does not adequately model this radiation problem, the NASA team felt that a conservative uncertainty of ±60% on radiative heating levels was required. Clearly, further research into radiative heat transfer would be of value to NASA's future planetary exploration endeavors. More details of the challenge of modeling CN radiation are discussed in Olejniczak (2003), Takashima (2003), Wright (2005), and Bose (2005) papers, which were authored by members of the NESC review team.

Another technical challenge presented by the Huygens mission was the transition to turbulence and the resulting turbulent convective heating levels. In general, the question of when boundary-layer transition occurs is very difficult to resolve and it is configuration (vehicle shape and size), destination (planetary atmosphere), and mission design dependent. The ESA concluded that transition would not occur until well after the peak heating point on the trajectory and, therefore, that the resulting turbulent heating augmentation would not be a significant design consideration. The NASA team disagreed with ESA conclusions regarding turbulence for two reasons:

NASA Engineering and Safety Center
Technical Assessment Report

Document #:
RP-05-67

Version:
1.0

Title:
Independent Technical Assessment of Cassini/Huygens Probe Entry, Descent and Landing (EDL) at Titan

Page #:
26 of 116

1. The fact that the heat shield would likely experience considerable ablation mass loss, which would have the effect of promoting transition.

2. The heat shield was not a continuous piece, but was comprised of individual tiles. Steps or gaps between these tiles could also promote transition.

However, without having the time to thoroughly study this transition problem, the NESC team could not provide a precise prediction of when transition would occur, but chose instead to adopt a conservative transition onset criterion based on the ratio of local boundary layer momentum thickness Reynolds number to local edge Mach number (Re_θ/M_e) = 200 (this is the working value being used for the Mars Science Laboratory development). The conclusion is that a detailed atmosphere, configuration and mission design dependent analysis of transitional/ turbulent behavior should be a requirement in the design of future interplanetary entry vehicles.

As noted previously, radiative heating at Titan is predominately due to the CN molecule. The amount of CN formed by the Huygens bow-shock is directly related to the amount of methane, CH_4, present in Titan's atmosphere. Initial estimates of CH_4 concentration (before the arrival of Cassini at Saturn) were as much as 5% by volume. These estimates were updated based on measurements made by Cassini during the T0 and TA passes at Titan to 1.8±0.5%. If the initial estimates were correct, then Huygens would definitely have encountered a more severe radiative heating environment, and may possibly have experienced heat shield failure. If future Titan missions are to be considered, more accurate knowledge of the CH_4 concentration in the atmosphere can be used to reduce design uncertainty margins. New information on CH_4 concentration should be available following post-flight analysis of the Huygens atmospheric science instruments.

A final technical lesson learned was the need for correct modeling of diffusion of different chemical species within the Huygens flow field. In an atmosphere such as Titan's, where species with a wide range of masses are formed (e.g. H atom with an atomic weight of 1 compared to the N_2 molecule with an atomic weight of 28), a rigorous, multi-component species diffusion model must be employed (See Appendix G; Sutton, 1998 or Ramshaw, 1990) rather than a simpler, binary-diffusion model which would be sufficient for the Earth's atmosphere. Such a model is part of both the DPLR and LAURA codes employed by the NASA team, but was not included in the original 2004 ESA analysis. Failure to include a multi-component diffusion model can result in convective heat rate prediction errors of up to 40%.

10.3 Analysis Process

Under ideal circumstances, the NESC review of the Huygens probe EDL should have been initiated at least one year before the probe release decision date, rather than less than 4 months

NASA Engineering and Safety Center
Technical Assessment Report

Document #:
RP-05-67

Version:
1.0

Title:
Independent Technical Assessment of Cassini/Huygens Probe Entry, Descent and Landing (EDL) at Titan

Page #:
27 of 116

before this date, as actually took place. That completion of this analysis was even possible owes to a number of fortuitous circumstances:

1. A multi-center team of researchers conducted, through NASA's ISP Program, a detailed technical-level systems analysis of an aerocapture mission to Titan in 2002-2003 (Lockwood, 2003). This study uncovered many of the aerothermodynamic technical challenges discussed in the previous section and, while not fully resolving them, had at least made progress in the development of the computational methods required for analysis of entry into Titan's atmosphere. This aerocapture analysis also included the development of the atmosphere model, which provided the composition, density, and wind profiles. Also, a detailed aerocapture flight simulation had been developed with Monte Carlo capability.

2. Additional aerothermodynamic model development had also been funded by the ISP Program through a series of competed aeroshell development proposals (See Appendix G; Wright, 2005; Bose, 2005).

3. The multi-center team that conducted this ISP Titan aerocapture systems study, while only marginally funded in 2004, still remained in close contact. The fact that the same team was used minimized the normal start-up issues.

4. The MRR team, established by JPL, was composed of researchers that were well-versed in the capabilities of the NESC team. This prevented any conflicts and allowed the talents of both groups to be used synergistically.

5. The ESA directly requested the assistance of NASA Ames during their delta-FAR in January 2004 in the areas of TPS and aerothermal modeling. This assistance was funded at a low level by the ISP Program until the NESC activity began. The results of this activity were used by the NASA team as the starting point and thus eliminated the normal time delays.

Several conclusions are offered based on the above facts. First, the fact the Huygens review could be successfully completed in the allotted time should not be taken as evidence that a similar future review could be completed in the same time span. Second, the state-of-the-art tools and methods available to NASA, particularly in aeroheating were, at best, only marginally sufficient to conduct this review. The need for future reviews of this nature should be identified and initiated early enough to allow sufficient time to complete the required analysis.

Another lesson learned is the importance of establishing clear links through which technical communication may be conducted. Unfortunately, communications at the technical level between the ESA and the NESC team were not satisfactory. In fact, there were no official communications until the November 4-5, 2004 technical interchange meeting (TIM) at the ESA Headquarters (based in Paris) (for aerothermodynamics), and the November 9-11, 2004 meeting at the ESTEC for the remaining disciplines. A partial communications link did exist between

NASA Ames and the ESA through an agreement to collaborate on an analysis of the Huygens probe that was initiated prior to the NESC study. Note that the ESA provided documents and answered clarification questions through e-mail prior to these meetings, but direct communication was not allowed. These communication difficulties were a result of the ESA team high workload as the time for probe release was nearing, the relatively small ESA team working Huygens a decade after the design phase, and the time-consuming requirements required dealing with the International Traffic in Arms Regulations (ITAR).

Another lesson learned was the importance of multiple independent, but cooperative, analysis capabilities in mission critical areas such as aeroheating analysis. This collaborative effort, which included two computational groups (LaRC and ARC), determined the aeroheating environment using two independent sets of computational tools (LaRC's LAURA and RADEQUIL codes, and ARC's DPLR and NEQAIR codes). Technical issues were identified by each group (both during this Huygens review and the aforementioned Titan aerocapture system analysis study). Collaborative efforts were undertaken to resolve these issues and reach a consensus viewpoint as to the best estimate of the entry heating environment. ESA was initially resistant toward adopting the aerothermodynamic methodology proposed by the NESC team. However, during the November 4-5, 2004 TIM, all of the NESC team's recommendations were accepted by the ESA, in large part because an agreement on these recommendations was reached by the two computational groups from NASA.

The final lesson learned is that NASA would benefit from a sustained funding level targeting advancement of critical computational tools to provide specific analysis capability for planned exploration spacecraft. Note that at present, there is little viable funding for the continued development of any of the tools used for the Huygens study. This will become a serious issue as NASA expands its exploration program from the small probes at Mars. For instance, any Venus probe or outer planets probes could experience radiative heating (different mechanism than seen at Titan), and large (human size) spacecraft could experience radiative heating at Mars and Earth. Also atmospheric modeling needs to be improved to capture more dynamics as the probes mature from unguided, uncontrolled vehicles to autonomous guidance and control for precise trajectory control and precision landing. The flight mechanics and simulation capability required to analyze these more capable vehicles must also be developed

Reference

Lockwood, M.K., *"Titan Aerocapture Systems Analysis"*, NASA LaRC, AIAA Paper 2003-4799.

Volume II: Appendices

A   ITA/I Request Form (NESC-FM-03-002)
B   Huygens Aerodynamic Database
C   Huygens Parachute System Evaluation
D   Atmospheric Properties
E   Flight Simulation
F   Monte Carlo Analyses and Results
G   Aerothermodynamics
H   Thermal Protection System

Appendix A. ITA/I Request Form (NESC-FM-03-002)

NASA Engineering and Safety Center Request Form		
Submit this ITA/I Request, with associated artifacts attached, to: **nrbexecsec@nasa.gov**, or to NRB Executive Secretary, M/S 105, NASA Langley Research Center, Hampton, VA 23681		
Section 1: NESC Review Board (NRB) Executive Secretary Record of Receipt		
Received (mm/dd/yyyy h:mm am/pm) 8/11/2004 12:00 AM	Status: New	Reference #: 04-069-I
Initiator Name: David Leckrone	E-mail: David.Leckrone@nasa.gov	Center: GSFC
Phone: (301)-286-5975, Ext ____	Mail Stop: 440	
Short Title: Independent Technical Assessment of Cassini/Huygens Entry, Descent and Landing		
Description: NESC Flight Sciences SPRT is requested to conduct a full, independent analysis of the Huygens EDL onto Titan. Flight data from other NASA missions should be utilized as appropriate. The NESC EDL solution will be provided to the Cassini Project and to ESA for comparison to the solution currently being adopted. Follow-up iterations of the computations and analysis by the FS/SPRT should be carried out as necessary, to assure the most accurate information possible is available to the Cassini/Huygens project to maximize the probability of mission success.		
Source (e.g. email, phone call, posted on web): email		
Type of Request: Independent Technical Assessment		
Proposed Need Date: September 11, 2004		
Date forwarded to Systems Engineering Office (SEO): (mm/dd/yyyy h:mm am/pm):		
Section 2: Systems Engineering Office Screening		
Section 2.1 Potential ITA/I Identification		
Received by SEO: (mm/dd/yyyy h:mm am/pm): 8/12/2004 12:00 AM		
Potential ITA/I candidate? ☒Yes ☐ No		
Assigned Initial Evaluator (IE): Approved to proceed at 8/12/04 NRB, Labbe to lead		
Date assigned (mm/dd/yyyy): 8/12/2004		
Due date for ITA/I Screening (mm/dd/yyyy):		
Section 2.2 Non-ITA/I Action		
Requires additional NESC action (non-ITA/I)? ☐Yes ☐ No		
If yes:		
Description of action:		
Actionee:		
Is follow-up required? ☐Yes ☐ No If yes: Due Date:		
Follow-up status/date:		
If no:		
NESC Director Concurrence (signature):		
Request closure date:		

Section 3: Initial Evaluation
Received by IE: (mm/dd/yyyy h:mm am/pm):
Screening complete date:
Valid ITA/I candidate? ☐ Yes ☐ No
Initial Evaluation Report #: NESC-PN-
Target NRB Review Date:
Section 4: NRB Review and Disposition of NCE Response Report
ITA/I Approved: ☐ Yes ☐ No | Date Approved: | Priority: - Select -
ITA/I Lead: , Phone (   )   -   , x
Section 5: ITA/I Lead Planning, Conduct, and Reporting
Plan Development Start Date:
ITA/I Plan # NESC-PL-
Plan Approval Date:
ITA/I Start Date | Planned: | Actual:
ITA/I Completed Date:
ITA/I Final Report #: NESC-PN-
ITA/I Briefing Package #: NESC-PN-
Follow-up Required? ☐ Yes ☐ No
Section 6: Follow-up
Date Findings Briefed to Customer:
Follow-up Accepted: ☐ Yes ☐ No
Follow-up Completed Date:
Follow-up Report #: NESC-RP-
Section 7: Disposition and Notification
Notification type: - Select - | Details:
Date of Notification:
Final Disposition: - Select -
Rationale for Disposition:
Close Out Review Date:

	NASA Engineering and Safety Center	Document #:	Version:
	**Technical Assessment Report**	RP-05-67	1.0

Title:	Page #:
Independent Technical Assessment of Cassini/Huygens Probe Entry, Descent and Landing (EDL) at Titan	32 of 116

Appendix B. Huygens Aerodynamic Database
Prepared by
Naruhisa Takashima
NASA LaRC

B.0   Introduction

A high-fidelity 6 degree-of-freedom (6DOF) aerodynamic database was developed to support the simulations of the Huygens probe from the Titan atmospheric interface to parachute deployment. Due to similarity in the forebody shape of the Huygens (60-degree sphere cone geometry) to that of the Genesis Sample Return Capsule (SRC), the Genesis aerodynamic database was used as the foundation for the Huygens aerodynamic database. The Genesis aerodynamic database, which in turn was based on that of the similar Stardust entry capsule, was constructed using data from engineering analysis tools, high-fidelity numerical analysis solutions (i.e., CFD), ground-based wind tunnel tests, and free-flight ballistic range tests. The details of the Stardust aerodynamic database are described by Mitcheltree et al.[1] and the details of the Genesis aerodynamic database are described by Desai et al.[2]

During its descent through the Titan atmosphere, the Huygens probe encountered several different types of flow fields from hypersonic flight in rarefied, transitional, and continuum regimes, to supersonic, transonic, and subsonic flight in the continuum regime. At the atmospheric interface, the flow field was highly rarefied and free molecular analysis was used to calculate the aerodynamics. As the probe descended and passed through the transitional regime, the flow became continuum and the aerodynamics were calculated using the modified-Newtonian theory. Typically, as was done for the Stardust and Genesis aerodynamic databases, Direct Simulation Monte Carlo (DSMC) analysis was used to calculate the aerodynamics in the transitional regime. However, for the present aerodynamic database, a bridging function in the Genesis aerodynamic database was used to model transitional aerodynamics and to connect the free molecular and continuum data in place of a detailed DSMC analysis due to schedule constraints. For the supersonic flow regime, the aerodynamics were based on existing ballistic range test data; the aerodynamics in the transonic and subsonic regimes were extrapolated from the supersonic data. Hence, the database was likely to be inaccurate in these flow regimes. However, since the Huygens probe was designed to deploy its parachute at Mach 1.5, these inaccuracies were deemed to be acceptable. The continuum hypersonic aerodynamics were constructed using extrapolation from the Genesis aerodynamic database anchored by solutions from the CFD code LAURA.

Although the Huygens probe and Genesis capsule have similar 60-degree sphere cone forebodies, the Huygens probe has a larger nose radius relative to the probe diameter, as shown

	NASA Engineering and Safety Center	Document #:	Version:
	Technical Assessment Report	RP-05-67	1.0

Title:	Page #:
Independent Technical Assessment of Cassini/Huygens Probe Entry, Descent and Landing (EDL) at Titan	33 of 116

in Figure B-1. The difference in the nose radius results in an axial force coefficient (CA) value that is greater than that of the Genesis capsule throughout the Mach range. At the hypersonic continuum limit, that difference is 6.8%. While the forebodies are similar, the difference in aftbody shapes and the resultant effects on the dynamic stability at supersonic conditions are substantial. Both geometries use truncated bi-conic shape for their aftbody shape, but for the Huygens probe, the aftbody maximum diameter is smaller than that of the forebody and the aftbody is tucked behind the forebody heat shield. By tucking the aftbody behind the heat shield, the axial center of gravity location is moved forward, which minimizes the effects of the aftbody on the wake flow and increases the dynamic stability at supersonic conditions.

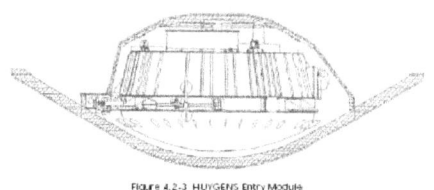

Figure 4.2-3 HUYGENS Entry Module

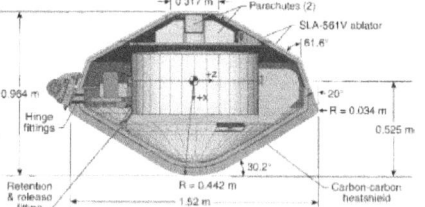

Fig. 2 Genesis SRC configuration.

- Huygens
  - 59.8 deg sphere cone
  - Diameter (D)= 2.7 m
  - Height = .985 m
  - Forebody Height= .62 m
  - Nose Radius (Rn)= 1.25m
  - Shoulder Radius= 0.05 m
  - Mass= 320 kg
  - $X_{cg}/D = 0.18$

- Genesis
  - 60 deg sphere cone
  - Diameter (D)= 1.52 m
  - Height = .964 m
  - Forebody Height= N/A
  - Nose Radius (Rn)= 0.442 m
  - Shoulder Radius= 0.034 m
  - Mass= 225 kg
  - $X_{cg}/D = 0.345$

Figure B-1. Huygens and Genesis Sample Return Capsule Configurations

B.1 Aerodynamic Database Modifications

To reflect the differences in the geometry between the Huygens probe and the Genesis SRC, the Genesis aerodynamic database was revised for certain flow regimes. To anchor the rarefied and continuum aerodynamics, free-molecular and modified-Newtonian analyses were performed using the DACfree engineering analysis tool. Figure B-2 shows how the probe geometry was simplified for the analyses. The results from the analyses were used to replace the free molecular and the hypersonic continuum limits data in the Genesis aerodynamic database. The existing Genesis bridging function with the new Huygens data was used to characterize the aerodynamics in the transitional flow regime.

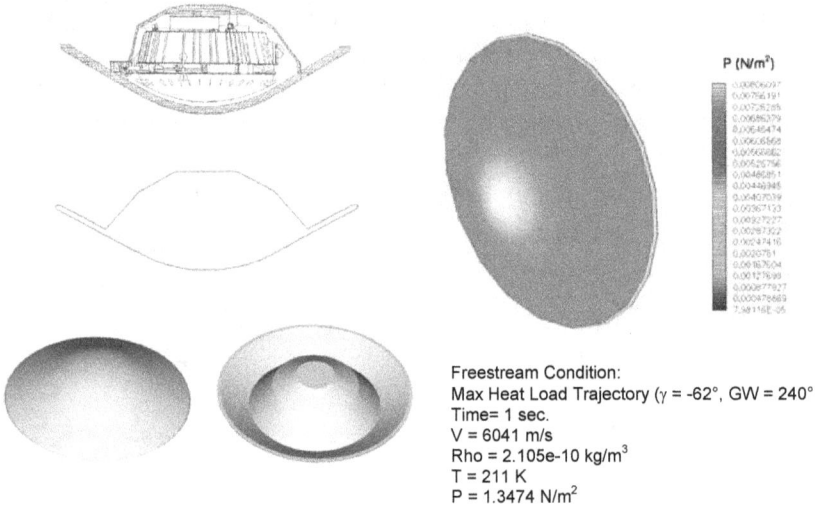

P (N/m²)

Freestream Condition:
Max Heat Load Trajectory (γ = -62°, GW = 240°
Time= 1 sec.
V = 6041 m/s
Rho = 2.105e-10 kg/m³
T = 211 K
P = 1.3474 N/m²

Figure B-2. Free Molecular/Modified Newtonian Analyses

For the supersonic flow regime, data from free flight tests of Huygens probe model, conducted at the Eglin Aeroballistic Range Facility (ARF) during September 1994[3], were used to characterize the aerodynamics. Based on the ballistic range test data, three versions of the Huygens aerodynamic database were generated. In the first version, the aerodynamics is characterized by curve-fits to a set of published data[4]. The second version is based on data reduction performed by Wayne Hathaway of Arrow Tech Associates. The third version is based on data reduction performed by Leslie Yates of Aerospace Computing, Inc. Figure B-3 presents the CA for the nominal attitude along a nominal reference trajectory for each of the three databases. This figure also shows the CA values in the Huygens aerodynamic database provided by the ESA as well as from LAURA CFD calculations (performed on the forebody-only geometry).

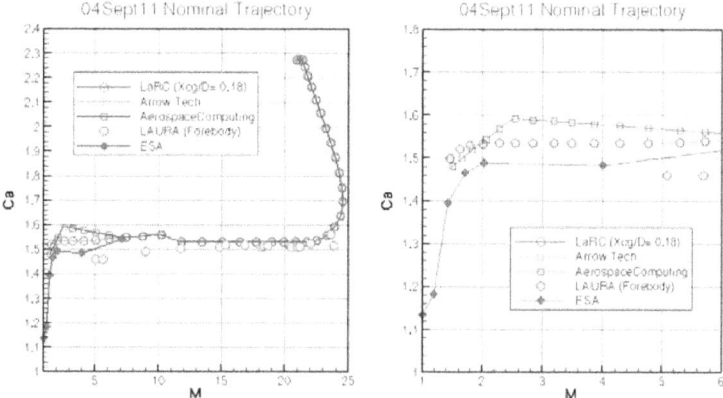

Figure B-3. Nominal CA variation with Mach Number

The comparison shows that the difference in the CA values for all versions of the aerodynamic database is small (within 10%). In contrast, Figure B-4 shows the difference in pitch damping coefficient (Cmq) with respect to attitude and Mach number being significant. The quantity Cmq indicates the dynamic stability of the probe in the pitch plane. The amplitude of the pitching motion increases when the Cmq value is larger than the stability limit and decreases when it is less. For the Huygens probe, all three aerodynamic databases show that the probe is dynamically unstable and that the amplitude of the motion tends to grow at small angles. However, because Cmq decreases with increasing angles-of-attack, the value of Cmq eventually becomes less than the stability limit, the probe becomes dynamically stable, and the amplitude of the motion tends toward a limit-cycle for most cases. Figure B-4 also shows that the probe is dynamically stable at Mach 2 according to all three versions of the database. However, the version of the aerodynamic database based on Arrow Tech analysis indicates that the probe is dynamically unstable at the parachute deploy condition of Mach 1.5.

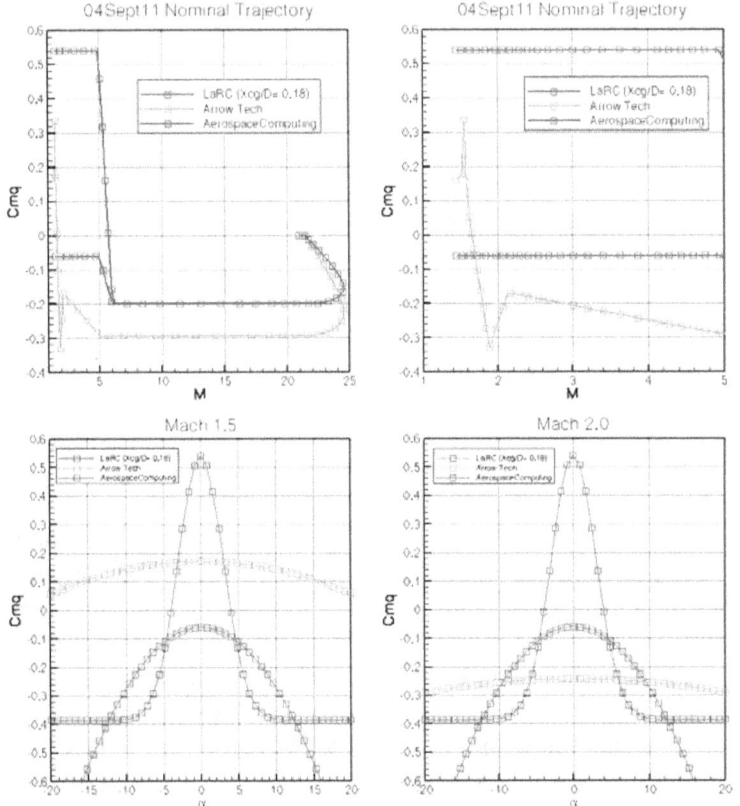

Figure B-4. Pitch Damping Coefficients

B.2   Uncertainties

In Table B-1, values listed under the heading "LaRC (Huygens)" illustrate the uncertainties that are assigned to all three aerodynamic databases. For comparison, the uncertainties for previous flight missions such as MER and Genesis, as well as uncertainties used by ESA, are listed. The present uncertainties are based on those used for MER, except for the values associated with the damping coefficient. An adder 3-sigma uncertainty of 0.15 is used to reflect the differences in the available data.

Table B-1. Aerodynamic Uncertainties

	LaRC (Huygens)	ESA	MER	Genesis
CA - Free molecular	5%	5%	5%	10%
Hypersonic	5%	5%	5%	4%
Supersonic	10%	5%	10%	10%
Subsonic		5%		5%
CN - Free molecular	0.01	10%	0.01	8%
Hypersonic	0.01	10%	0.01	8%
Supersonic	0.01	10%	0.01	5%
CY - Free molecular			0.01	8%
Hypersonic			0.01	8%
Supersonic			0.01	5%
Cm - Free molecular	0.005	10%	0.005	12%
Hypersonic	0.003	10%	0.003	10%
Supersonic	0.005	10%	0.005	8%
Cn - Free molecular			0.005	12%
Hypersonic			0.003	10%
Subsonic			0.005	8%
Cmq & Cnr - Free molecular	0.15		0.090	
Hypersonic	0.15	5%**	0.090	0.28*
Supersonic	0.15	10%***	[+100% to -50%] + [0 to 0.1]*	0.2*
*Uniform distribution		**M>1.7	***1.0 < M < 1.4	

In Figure B-5, the nominal CA variation with Mach is shown. The ESA uncertainty is also shown to illustrate the relative differences. The figure shows that for Mach numbers greater than five, all four sets of data fall within the uncertainties. From Mach numbers of 2.2 to 4.8, the CA values based on Aerospace Computing are slightly greater than the rest due to the choice of parametric values used to bridge the supersonic and hypersonic data, but overall the differences are small (again within 10%).

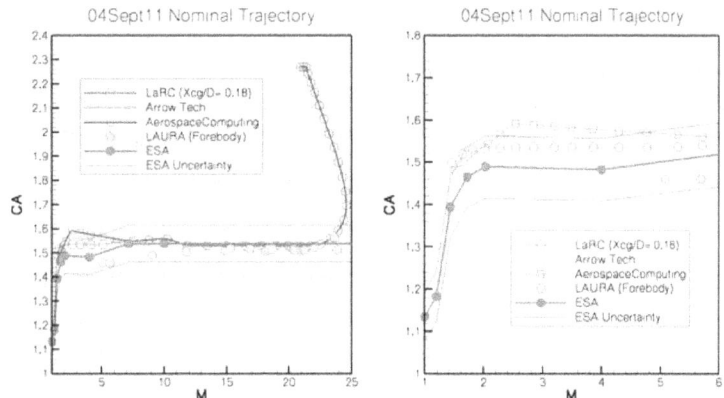

Figure B-5. Nominal CA vs. Mach Number with ESA Uncertainty

**NASA Engineering and Safety Center
Technical Assessment Report**

Document #:	Version:
RP-05-67	1.0

Title:

**Independent Technical Assessment of Cassini/Huygens
Probe Entry, Descent and Landing (EDL) at Titan**

Page #:
38 of 116

Figure B-6 shows the pitch damping coefficients with uncertainties. The uncertainty for the ESA data is relatively small compared to the uncertainties used for MER. The uncertainty value assigned for the damping coefficient for the present aerodynamic databases are slightly less conservative than the MER uncertainty, which is a compromise between the MER uncertainty and the ESA uncertainty.

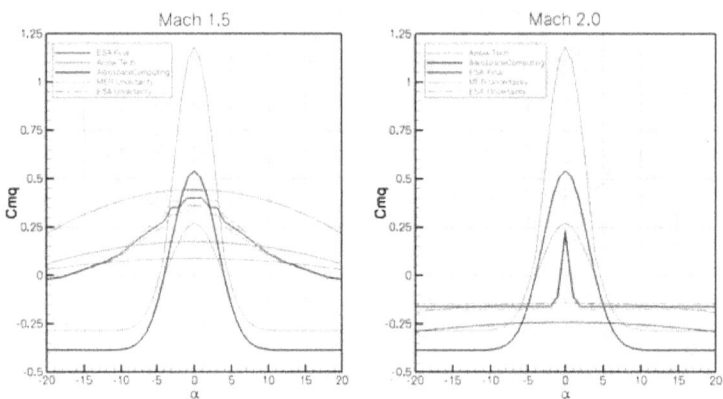

Figure B-6. Pitch Damping Coefficients with Uncertainties

B.3   Summary

Three different aerodynamic databases for the Huygens probe were generated for the Huygens probe EDL analyses. The databases were based on the Genesis SRC aerodynamic database with updates to reflect the differences in the Huygens geometry, particularly the aftbody shape. Three independent analyses of a single set of ballistic range test data were used to construct the three databases. Relatively conservative values of uncertainties that are consistent with heritage values were assigned to the database for use in Monte Carlo simulation analyses.

References

[1]Mitcheltree, R. A. et al., "Aerodynamics of Stardust Sample Return Capsule", *Journal of Spacecraft and Rocket*, Vol. 36, No. 3, 1999, pp. 463-469.

[2]Desai, P. N. and Cheatwood, F. M., "Entry Dispersion Analysis for the Genesis Sample Return Capsule," *Journal of Spacecraft and Rockets*, Vol. 38, No. 3, 2001, pp.345-350.

[3]Winchenbach, G. et al., "Huygens Probe Aerodynamics Free Flight Test Results: ARF Ballistic Spark Range," WL-ARO-95-0022, March 1995.

[4]Chapman, G., T. and Yates, L. A., "Dynamics of Planetary Probes: Design and Testing Issues," AIAA Paper 1998-797, Jan. 1998.

Appendix C. Huygens Parachute System Evaluation

Prepared by
Juan R. Cruz and Scott A. Striepe
Exploration Systems Engineering Branch
NASA Langley Research Center

C.0   Introduction

An independent evaluation of the parachute system for the Huygens probe was conducted. Because of time constraints, only two issues were evaluated in detail. These issues were:

1. **The opening loads and structural strength of the pilot and main parachutes (including their attachments to the probe).** Concerns had been raised regarding this issue because the envelope of possible deployment conditions was different than those to which the parachute system was designed, built, and qualified.

2. **The descent time of the probe from pilot parachute deployment to landing.** Descent time is strongly dependent on the drag coefficient of the drogue parachute and thus places emphasis on its drag modeling. Mission requirements specified a descent time from 2.0 to 2.5 hours in order to maintain communication with Cassini.

Both of these issues are heavily dependent on the drag models for the parachute. Thus, a large portion of the effort involved in this evaluation centered on the generation of parachute drag models, including uncertainty limits and distributions.

Other issues regarding the Huygens parachute system were evaluated by reviewing the published literature on the system development (refs. 1 to 7) and additional documentation provided by the ESA and its contractors (refs. 8 to 10). This evaluation of the published literature and documentation found no other issues of significant concern. All work performed by the ESA and its contractors regarding the Huygens probe parachute system was found to be of high quality and thoroughly documented. The remainder of this Appendix is divided in three sections.

C.1   The parachutes drag models are discussed, given that they play a key role in the parachute opening loads and probe descent time calculations.

C.2   Analyses and results for the opening loads and structural strength of the pilot and main parachute are discussed.

C.3   The descent time calculations and results are presented.

NASA Engineering and Safety Center Technical Assessment Report	Document #: RP-05-67	Version: 1.0
Title: **Independent Technical Assessment of Cassini/Huygens Probe Entry, Descent and Landing (EDL) at Titan**		Page #: 40 of 116

The help and information provided by Dr. J. Stephen Lingard of Vorticity Systems through e-mails and phone conversations is gratefully acknowledged.

C.1   Parachutes Drag Models

The Huygens parachute system (referred to in the Huygens literature as the Descent Control Subsystem (DCSS)) consists of three Disk-Gap-Band (DGB) parachutes: pilot, main, and stabilizer (i.e., drogue). In Figure C-1, the Huygens entry, descent, and landing (EDL) sequence is shown, emphasizing the operation of the parachute system. Key parameters for the three parachutes are provided in Table C-1. Because the design and geometric porosity of the main and stabilizer parachutes are the same, and their trailing distances from the probe are relatively large, they can use the same drag coefficient (C_{D0}) model. Thus, a complete *set* of parachute drag models consists of two drag models – one for the pilot parachute and one for the main and stabilizer parachutes.

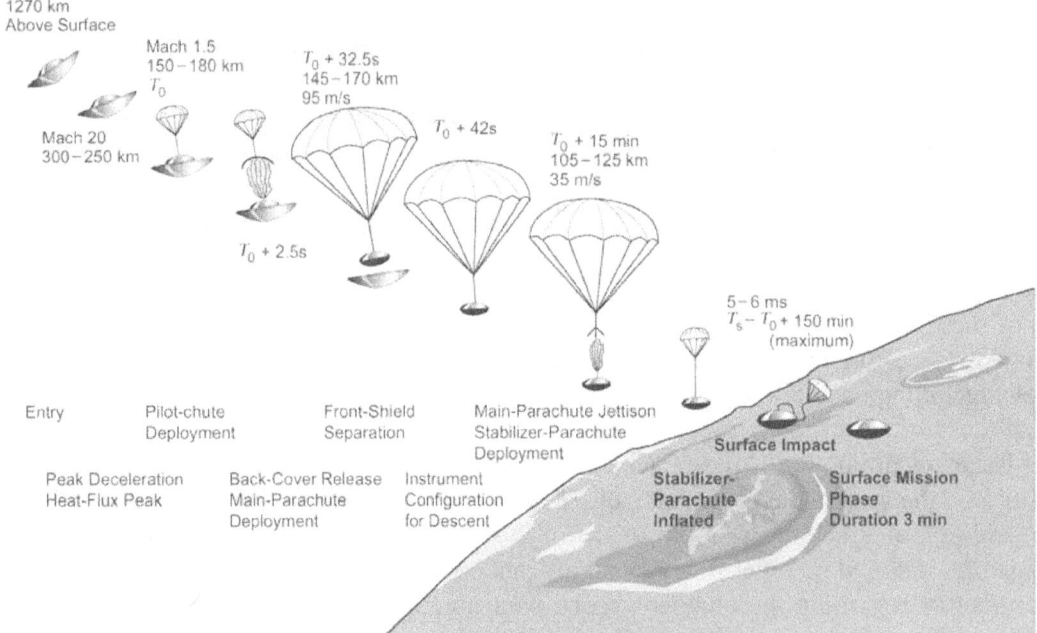

Figure C-1. Huygens EDL Sequence
(Graphic from reference 11)

NASA Engineering and Safety Center
Technical Assessment Report

Document #:	Version:
RP-05-67	1.0

Title:

Independent Technical Assessment of Cassini/Huygens Probe Entry, Descent and Landing (EDL) at Titan

Page #:

41 of 116

Table C-1. Key Parameters for the Huygens Parachutes

Parachute	Nominal Diameter D_0	Nominal Area S_0	Geometric Porosity λ_g
Pilot	2.59 m	5.27 m^2	13.1%
Main	8.30 m	54.1 m^2	22.4%
Stabilizer (drogue)	3.03 m	7.21 m^2	22.4%

Five parachute drag model sets were used in the present evaluation. The first drag model set, denoted here as the ESA model, is given in reference 8[3]. This is the baseline drag model set used by ESA in its trajectory calculations. It was derived from Huygens-specific wind tunnel tests. This drag model set has the tightest uncertainty band of all the model sets: ±5% of the mean at the 3σ uncertainty level with a normal uncertainty distribution. In the ESA model, the drag coefficients depend on Mach number, M, and Reynolds number per meter, Re^*.

The second parachute drag model set, denoted here as the NASA 1 model, was created by adjusting a Viking-heritage drag model currently in use at NASA to the geometric porosity of the Huygens parachutes[4]. In general, this drag model set has the widest uncertainty band of all the drag model sets. The purpose of creating this drag model set was to have a completely independent estimate of the Huygens parachutes drag models. In the NASA 1 model, the drag coefficients depend only on the Mach number, M.

The third parachute drag model set, denoted here as the NASA 2 model, was derived from the ESA model combined with a re-analysis of the Huygens parachute system drop test data performed by Vorticity (refs. 6 and 10). In this drag model set, the original mean value of C_{D0} from the ESA model for the pilot parachute was multiplied by 1.11, and the uncertainty band defined as ±10% of the new mean value with a triangular uncertainty distribution. The original mean value of C_{D0} in the ESA model for the main and stabilizer parachutes was multiplied by 1.22, and the uncertainty band defined as ±10% of the new mean value with a triangular uncertainty distribution. Increasing the drag coefficients was performed to reflect the higher values of C_{D0} implied by the Huygens parachute system drop test. In the NASA 2 model the drag coefficients depend on Mach number, M, and Reynolds number per meter, Re^*.

Vorticity generated two additional parachute drag model sets that took into account the re-analyzed data from the Huygens parachute system drop test (refs. 6 and 10). These drag models

---

[3] Dr. Lingard provided clarifications on this drag model through e-mail communications.

[4] The geometric porosity of the Viking parachute was 12.5%. Since the drag coefficient of a parachute is sensitive to the geometric porosity (increasing the geometric porosity decreases the drag coefficient), the drag coefficient of the Viking parachute drag model was adjusted for geometric porosity based on data presented in reference 3.

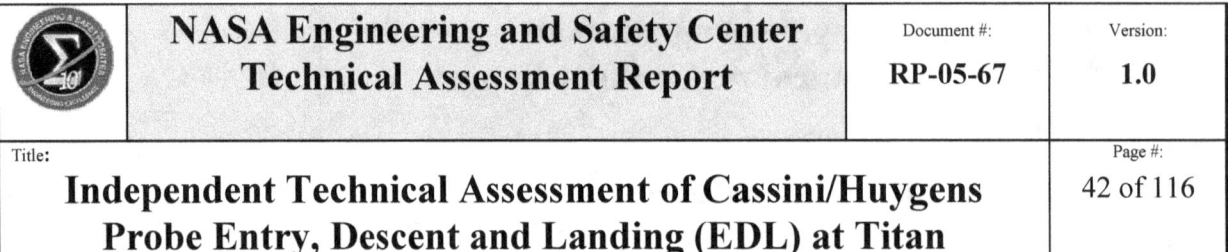

NASA Engineering and Safety Center Technical Assessment Report	Document #: RP-05-67	Version: 1.0

Title: Independent Technical Assessment of Cassini/Huygens Probe Entry, Descent and Landing (EDL) at Titan	Page #: 42 of 116

are denoted as Vorticity 1 and Vorticity 2 and were used by Vorticity to generate the descent time Monte Carlo results presented in reference 10. In these drag model sets, the mean values and uncertainties of C_{D0} from the Huygens parachute wind tunnel tests were slightly increased to better agree with the re-analyzed data from the Huygens parachute system drop tests (ref. 10).

A summary of these three parachute drag model sets is given in Table C-2.

Table C-2. Huygens Parachute Drag Model Sets

Model	Data Source(s)	Adjustments	Dependency
ESA	Huygens wind tunnel test data	None – used as is	M and Re^*
NASA 1	Viking heritage	Adjusted to match Huygens parachutes geometric porosities	M
NASA 2	Huygens wind tunnel test data and system drop test	Increased Huygens wind tunnel test data to match system drop test data	M and Re^*
Vorticity 1	Huygens wind tunnel test data and system drop test	Increased Huygens wind tunnel test data based on system drop test data	M and Re^*
Vorticity 2	Huygens wind tunnel test data and system drop test	Increased Huygens wind tunnel test data based on system drop test data	M and Re^*

C.2 Parachute Opening Loads and Structural Strength

Two sets of calculations were performed in support of the opening loads and structural strength evaluation of the pilot and main parachutes. First, an opening loads validation phase was conducted by performing a series of calculations to determine how the NASA opening loads model compared with those of ESA (as presented in reference 8) when using the same set of assumptions. These calculations were performed to determine if there were any issues in how the opening loads were being calculated – either by ESA or NASA. This set of calculations indicated that for the pilot parachute the ESA and NASA calculations differed by 1.1 percent at the highest load condition. For the main parachute, the ESA and NASA calculations differed by 8.0 percent at the highest loads conditions. These differences are within the uncertainty limits of the analyses, and thus provided confidence in both sets of calculations.

Having successfully completed the validation phase, a series of opening loads analyses were conducted with the NASA 1 drag model. This drag model was used because it was the most

NASA Engineering and Safety Center
Technical Assessment Report

Document #:	Version:
RP-05-67	1.0

Title:

Independent Technical Assessment of Cassini/Huygens Probe Entry, Descent and Landing (EDL) at Titan

Page #:
43 of 116

conservative (i.e., had the highest mean C_{D0} and uncertainty level at the conditions for pilot and main parachute deploy) of all the drag models being considered. A summary of the NASA opening loads calculations for the pilot and main parachute, and the strength of the weakest component of the system (from data provided by ESA), are provided in Table C-3. From the Monte Carlo results in Table C-3 it is evident that the opening loads exceed the system strength of the weakest component in less than 0.13 percent of cases. Considering the stacked, worst-case results, the opening loads exceed the system strength of the weakest component. However, given the extreme conservatism of this opening load calculation (stacked worst-case with a pilot parachute trigger malfunction), the risk of a structural failure due to the pilot or main parachute opening loads was deemed to be very low. Thus, risk to the mission due to structural failure associated with the pilot or main parachute opening was retired as a concern.

Table C-3. Pilot and Main Parachute Opening Loads and System Strength Summary

Parachute	ESA/Vorticity Analysis NewMaxMq Condition	NASA Monte Carlo Analysis 99.87 Percentile	NASA Stacked Worst Case Analysis NewMaxMq Condition	Weakest Component With Safety Factor But No Reserves
Pilot	2,336 N (ref. 8)	2,133 N	2,520 N	2,500 N Back Cover Ultimate Load (ref. 9)
Main	17,600 N (ref. 8)	15,512 N	20,112 N	≥ 17,600 N Swivel (ref. 8)

Notes:
1) All calculations are performed with the NASA 1 drag model.
2) For the NewMaxMq condition to happen, a pilot parachute trigger malfunction has to occur.

C.3 Descent Time Calculations and Results

The mission requirements specified a descent time from pilot parachute deployment (time T_0) to landing (time T_S) between 2.0 to 2.5 hours. As can be seen from Figure C-1, most of the descent time is spent under the stabilizing parachute. Thus, the drag model of the stabilizing parachute is critical in predicting the descent time.

Two final sets of descent time calculations were performed: one by Vorticity using the Vorticity 1 and 2 drag models (ref. 10), and another by NASA using the NASA 2 drag model. All these drag models were derived from the Huygens wind tunnel test data and system drop test data. The main difference between the Vorticity 1 and 2 drag models and the NASA 2 drag model arose from the emphasis placed on the wind tunnel test data versus that from the system drop test data. Vorticity chose to bias its drag models towards the wind tunnel test data. NASA chose to

NASA Engineering and Safety Center Technical Assessment Report	Document #: RP-05-67	Version: 1.0
Title: **Independent Technical Assessment of Cassini/Huygens Probe Entry, Descent and Landing (EDL) at Titan**	Page #: 44 of 116	

bias its drag models towards the system drop test data. Because the system drop test data yielded higher estimates of C_{D0}, the NASA 2 drag model for the stabilizing parachute had values of C_{D0} approximately 16 percent higher than those in the Vorticity 1 and 2 models. The higher values of C_{D0} used in the NASA 2 model yielded longer descent times than those calculated by using the Vorticity 1 and 2 models. Vorticity and NASA personnel discussed these differences in interpretation of the available drag data, but no agreement was reached as to which drag model was more accurate. Both Vorticity's and NASA's analysis results were provided to the ESA/NASA review team. The results of Vorticity's and NASA's descent time analyses are presented in Table C-4.

Table C-4. Summary of Final Descent Time Calculations

Drag Model	Atmosphere Model	Mean Descent Time	Monte Carlo Analyses with Descent Times Exceeding 2.5 hrs
Vorticity 1	Not available	Not Available	0.4%
Vorticity 2	Not available	Not Available	4%
NASA 2	Fminmax variation atmosphere model in Titan-GRAM V1.0	2.51 hr	55%
NASA 2	TA average profile mean atmosphere in Titan-GRAM V1.0	2.45 hr	24%

References

1) Lingard, J. S. and Underwood, J. C., Wind Tunnel Testing of Disk-Gap-Band Parachutes related to the Cassini-Huygens Mission, AIAA Paper 93-1200, 1993.

2) Lorenz, R. D., Scientific Implications of the Huygens Parachute System, AIAA Paper 93-1215, 1993.

3) Lingard, J. S. and Underwood, J. C., The Effect of Low Density Atmospheres on the Aerodynamic Coefficients of Parachutes, AIAA Paper 95-1556, 1995.

4) Neal, M. F. and Wellings, P. J., Design and Qualification of the Descent Control Subsystem for the Huygens Probe, AIAA Paper 95-1533, 1995.

5) Underwood, J. C., Development Testing of Disk-Gap-Band Parachutes for the Huygens Probe, AIAA Paper 95-1549, 1995.

NASA Engineering and Safety Center Technical Assessment Report	Document #: RP-05-67	Version: 1.0
Title: **Independent Technical Assessment of Cassini/Huygens Probe Entry, Descent and Landing (EDL) at Titan**		Page #: 45 of 116

6) Underwood, J. C., A System Drop Test of the Huygens Probe, AIAA Paper 97-1429, 1997.

7) Underwood, J. C. and Sinclair, R. J., Wind Tunnel Testing of Parachutes for the Huygens Probe, in: Proceedings – Wind Tunnels and Wind Tunnel Test Techniques, pp. 47.1 – 47.11, Royal Aeronautical Society, London, 1997.

8) Underwood, J. C., Huygens DCSS Study – Final Report, Vorticity Document VOR-RE-0301, Issue 3, November 26, 2003.

9) Lebleu, D., Entry and Descent Performance Report, Mission Analysis/Probe System Performance Report for Yelle and Gravity Wave Models, Alcatel Space Document HUY.ASP.HIT.RE.0005, Issue 02, Revision B, October 12, 2004.

10) Lingard, J. and Underwood, J., Huygens Recovery Mission – Descent Control Subsystem DFAR Review Descent Time Issue, Vorticity Presentation, November 2004.

11) Cassini-Huygens Saturn Arrival, Press Kit, National Aeronautics and Space Administration, June 2004.

Symbols

C_{D0}	parachute drag coefficient
D_0	parachute nominal diameter
M	Mach number
Re^*	Reynolds number per meter
S_0	parachute nominal area
T_S	time at landing
T_0	time at pilot parachute deployment
λ_g	parachute geometric porosity
σ	standard deviation

Acronyms

AIAA	American Institute of Aeronautics and Astronautics
DCSS	Descent Control Subsystem
DGB	Disk-Gap-Band
EDL	Entry, Descent, and Landing
ESA	European Space Agency
NASA	National Aeronautics and Space Administration

Appendix D. Atmospheric Properties

Prepared by
C. G. (Jere) Justus
MSFC/Morgan Research

D.0   Introduction

The Titan Global Reference Atmospheric Model (Titan-GRAM; Justus et al., 2004) was the atmospheric model used for this study. Initially, Titan-GRAM was based on minimum/average/maximum temperature and density envelopes prescribed by Roger Yelle et al. (1997). These envelopes, shown in Figure D-1, were used in the design and analysis of Huygens by the ESA team. The Yelle profiles were intended to account for Titan atmospheric changes due to expected seasonal and time-of-day variations (over the whole globe), expected latitudinal variations (over the whole globe), and effects of uncertainties in data from Voyager and other sources, on which the Yelle profiles were based.

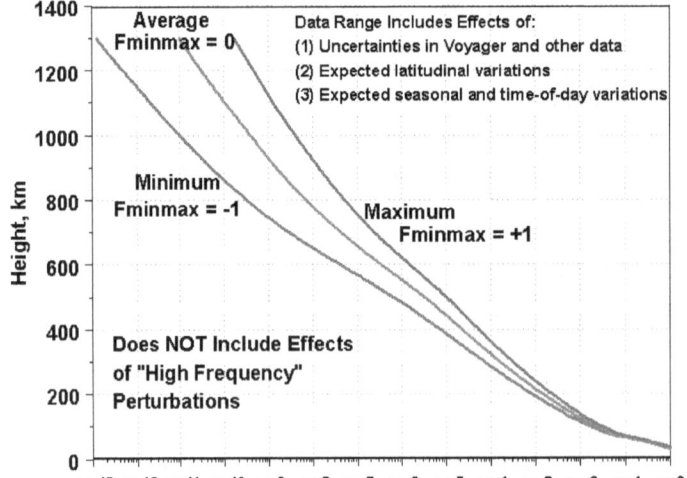

Figure D-1. Minimum/Average/Maximum Density Envelope

Profiles shown in Figure D-1 do not include the effects of "high frequency" variations, such as those caused by atmospheric gravity waves or turbulence. These are treated in Titan-GRAM (Justus et al., 2004) by a one-step Markov model, suitable for generating atmospheric

NASA Engineering and Safety Center Technical Assessment Report	Document #: RP-05-67	Version: 1.0
Title: **Independent Technical Assessment of Cassini/Huygens Probe Entry, Descent and Landing (EDL) at Titan**		Page #: 47 of 116

perturbations for use in the Monte Carlo simulations. An example Titan-GRAM perturbation model output is illustrated in Figure D-2. The Titan-GRAM model produces perturbations of similar magnitude to those of the gravity wave model by Strobel and Sicardy (1997). A major difference is that the Strobel and Sicardy model produces "monochromatic" wave perturbations, while perturbations from the Titan-GRAM model have a spectrum of various wavelength contributions (consistent with a "Dryden" spectrum, frequently used in turbulence simulation).

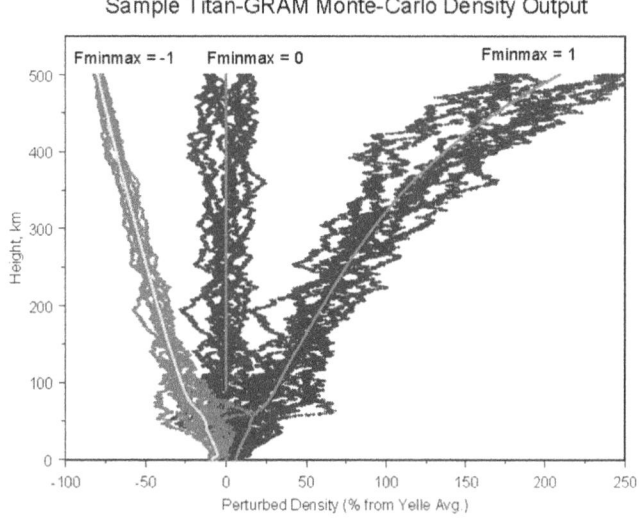

Figure D-2. Examples of Output from Titan-GRAM's "High Frequency" Perturbation Model

For this study, the following two new options were added to Titan-GRAM:

1. Use atmospheric data from the General Circulation Model (GCM) of Hourdin et al. (1995).

2. Read and use any "auxiliary profile" of temperature and density versus altitude.

In exercising either of these options, the alternate data (i.e., either the GCM data or data from the auxiliary profile) totally replaces the data from the original Yelle profiles. One example of an auxiliary profile, developed for potential use in this study, is temperature and density data from the Infrared Space Observatory (ISO; Coustenis et al., 2003). Figure D-3 compares density profiles from the original Yelle et al. (1997) data with results from the Hourdin GCM and Coustenis ISO profile. The GCM data in Figure D-3 were evaluated at the approximate location and time of the planned Huygens entry. The ISO profile is from disk-averaged observations made throughout calendar year 1997 and thus represents a wide range of locations and times. Up to a height of 500 km, both GCM and ISO density profiles are seen to be well within the envelope of minimum/maximum Yelle profiles.

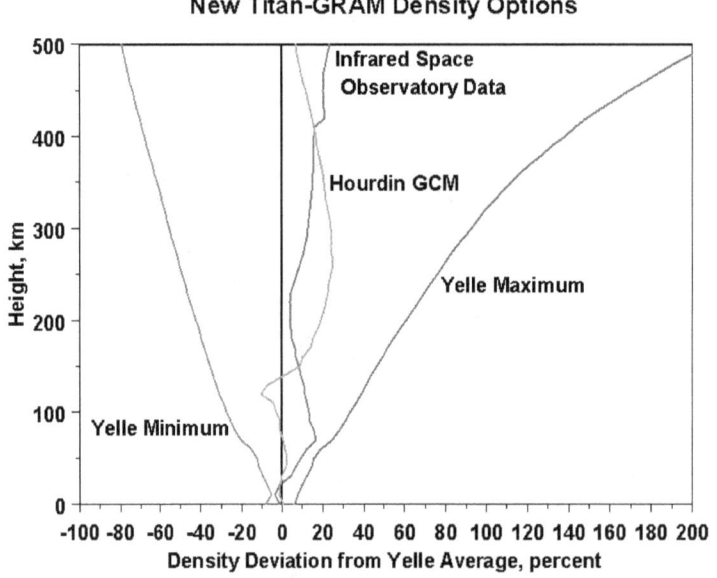

Figure D-3. Comparison of Density Profiles from Hourdin et al. (1995) GCM and ISO Profile (Coustenis et al, 2003) with original Yelle Density Profiles

During the period prior to Huygens Titan entry on January 14, 2005, the Cassini/Huygens mission plan included three opportunities for remote sensing of Titan's atmosphere by Cassini instruments during relatively close Titan flyby operations.

1. Titan flyby "T0" (July 3, 2004).
2. Titan flyby "TA" (November 15, 2004).
3. Titan flyby "TB" (December 13, 2004).

Titan atmospheric data results from the Titan flyby T0 were discussed at a workshop in Greenbelt, Maryland (Yelle, 2004). Refer to the Trip Report, provided at Attachment A of this Appendix, for a full summary of this workshop. T0 atmospheric data were made available in digital format shortly after the workshop by the Titan Atmospheric Model Working Group (TAMWG, Yelle, personal communication). Titan atmospheric data results from the Titan flyby TA were discussed at a workshop in Monrovia, California (Yelle, 2004). Refer to the Trip Report, provided at Attachment B of this Appendix, for a full summary of this workshop. TA

NASA Engineering and Safety Center Technical Assessment Report	Document #: RP-05-67	Version: 1.0
Title: **Independent Technical Assessment of Cassini/Huygens Probe Entry, Descent and Landing (EDL) at Titan**		Page #: 49 of 116

atmospheric data were made available in digital format shortly afterward by the TAMWG (Yelle, personal communication). There was insufficient time to fully process results from the TB flyby before Huygens entry. However, preliminary results indicated very little change from the TA results, so no further update was deemed necessary (James Frautnick, JPL, personal communication).

Figure D-4 compares the TAMWG T0 nominal density profile (0% Argon, 1.9% methane mole fraction) with the original Yelle et al. envelope. Up to an altitude of about 400 km, there is very good agreement with the original Yelle average profile. Near 800 km (well above altitudes of maximum concern for Huygens entry), the T0 nominal profile has density of about a factor of two higher than original average conditions. Figure D-4 also illustrates the wide range of density values contained within the original Yelle envelope. Densities at about 300 km are approximately a factor of two higher and lower (+100% to -50%) with respect to original Yelle average.

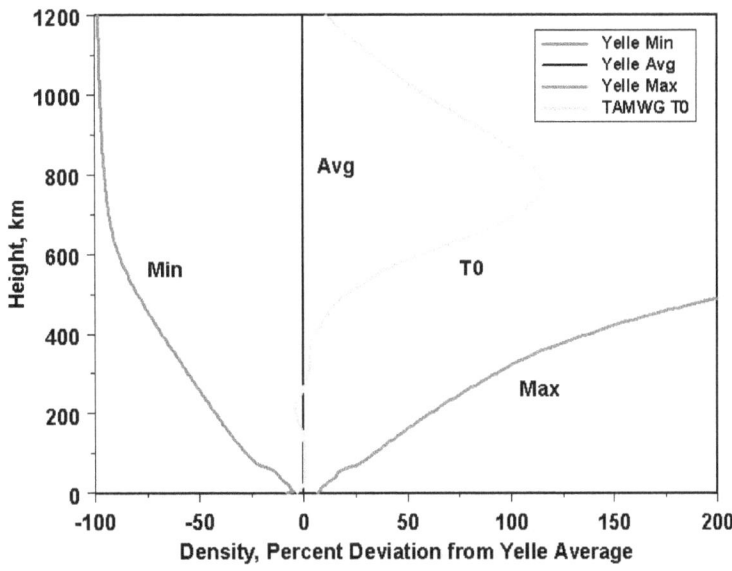

**Figure D-4. Comparison of Nominal (0% Argon, 1.9% methane mole fraction) TAMWG
T0 Density Profile (Yelle, 2004) with original Yelle et al. (1997) Density Profiles**

TAMWG T0 results include a range of assumed argon and methane values, as shown in Table D-1. Figure D-5 compares all of these profiles with the original Yelle envelope: all are very close to the original Yelle average density.

NASA Engineering and Safety Center Technical Assessment Report	Document #: RP-05-67	Version: 1.0
Title: **Independent Technical Assessment of Cassini/Huygens Probe Entry, Descent and Landing (EDL) at Titan**		Page #: 50 of 116

Table D-1. TAMWG October 6, 2004 Titan T0 Profiles and Mole Fractions of Molecular Nitrogen (N_2), Argon (Ar), and Methane (CH_4), in Percent

```
File Name          N₂%    CH₄%    Ar%      Description
---------------    -----  ------  ---      --------------------
HuygensT0_0.txt    98.1(-) 1.9(+)  0    Nominal CH₄, Low Ar
HuygensT0_1.txt    91.1(-) 1.9(+)  7    Nominal CH₄, High Ar
HuygensT0_2.txt    99.0(-) 1.0(+)  0    Low CH₄, Low Ar
HuygensT0_3.txt    92.0(-) 1.0(+)  7    Low CH₄, High Ar
HuygensT0_4.txt    97.0(-) 3.0(+)  0    High CH₄, Low Ar
HuygensT0_5.txt    90.0(-) 3.0(+)  7    High CH₄, High Ar
```

(-) means lower values in troposphere
(+) means higher values in troposphere

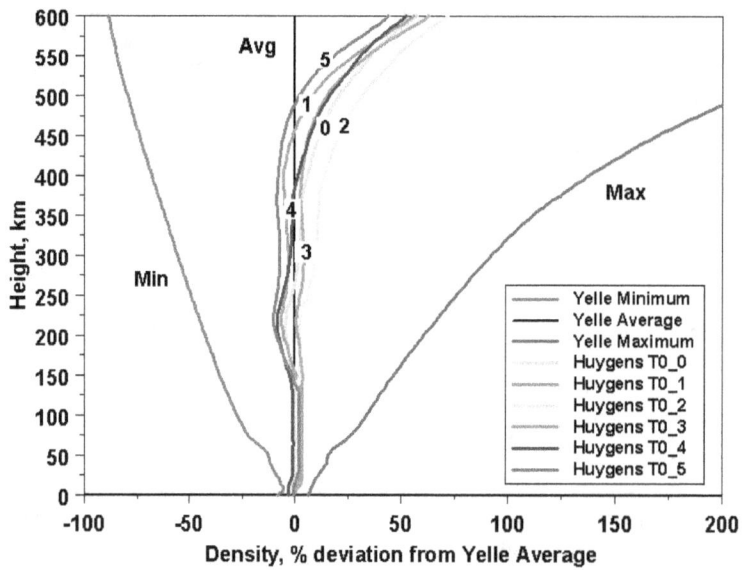

Figure D-5. Comparison of TAMWG T0 Density Profiles 0 through 5 (Yelle, 2004a) with Original Yelle et al. (1997) Density Profiles

Figure D-6 compares the TAMWG T0 nominal temperature profile (0% argon, 1.9% methane mole fraction) with the TAMWG TA maximum methane temperature profile (0% argon, 2.3% methane mole fraction). Up to an altitude of about 400 km, there is very good agreement between the two profiles. An "unofficial" estimate of the residual uncertainty in the T0 profile is also indicated (dashed lines) in Figure D-6. The "official" TAMWG estimate of residual uncertainty in the TA profile is also shown (solid lines). Corresponding T0 and TA density profiles (with uncertainty estimates) are provided in Figure D-7.

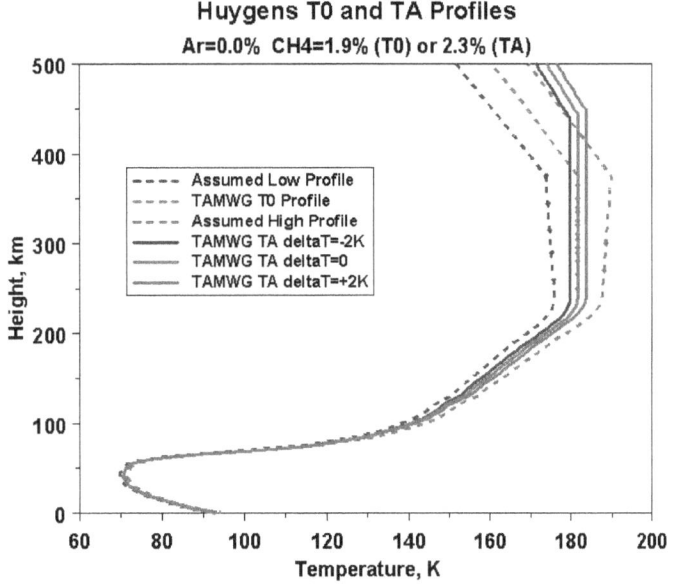

Figure D-6. Comparison of TAMWG T0 (nominal) and TA (maximum methane) Temperature Profiles

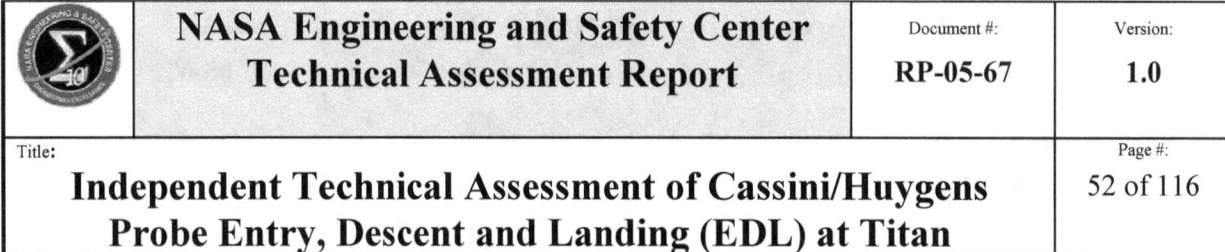

NASA Engineering and Safety Center Technical Assessment Report	Document #: RP-05-67	Version: 1.0
Title: Independent Technical Assessment of Cassini/Huygens Probe Entry, Descent and Landing (EDL) at Titan		Page #: 52 of 116

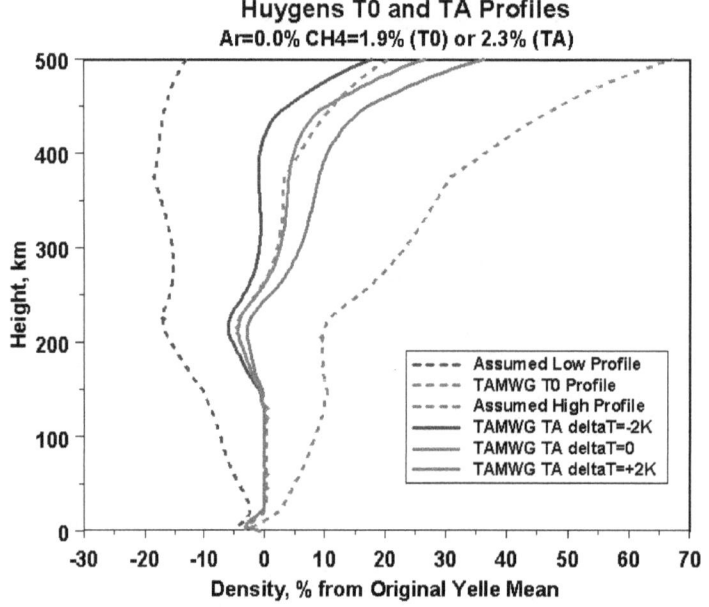

Figure D-7. Comparison of TAMWG T0 (nominal) and TA (maximum methane) Density Profiles (Yelle, 2004a) with original Yelle et al. (1997) Density Profiles

In addition to temperature and density, Titan-GRAM also provides a model for winds, both mean and perturbed. The TAMWG TA workshop endorsed Titan-GRAM winds as being consistent with the latest Cassini remote sensing wind estimates from T0 and TA Titan flyby operations, both in terms of mean wind profile and profile of wind perturbation standard deviations. Refer to Attachment B, TAMWG TA Trip Report, for more details.

NASA Engineering and Safety Center Technical Assessment Report	Document #: RP-05-67	Version: 1.0
Title: Independent Technical Assessment of Cassini/Huygens Probe Entry, Descent and Landing (EDL) at Titan		Page #: 53 of 116

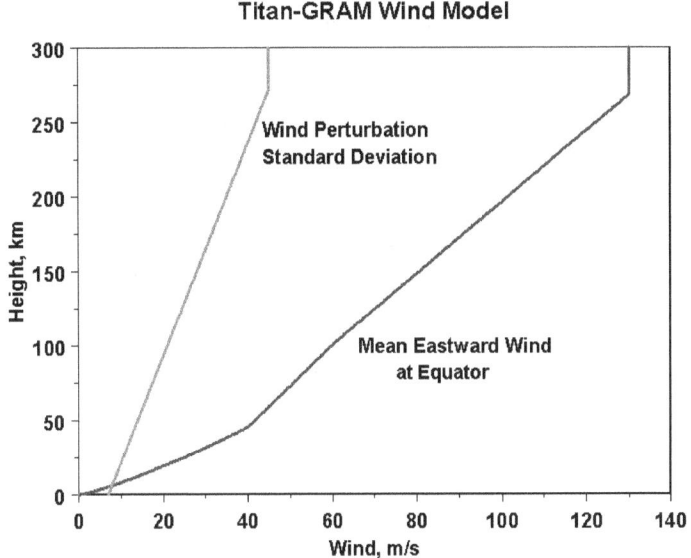

Figure D-8. Profiles of Equatorial Mean Eastward Wind and Wind Perturbation Standard Deviations, from the Titan-GRAM Model (Justus et al., 2004)

Tabular data from both the TAMWG T0 and TA temperature and density results were generated and formatted for use in Titan-GRAM as auxiliary profiles. This allowed Titan-GRAM to use, without program code changes, the latest atmospheric results for expected mean conditions, while retaining the same features of Titan-GRAM's "high frequency" perturbation model for use in Monte Carlo studies of guidance, system performance, and aeroheating of the Huygens probe.

References

Coustenis, Athena; Salama, A.; Schulz, B.; Ott, S.; Lellouch, E.; Encrenaz, T. H.; Gautier, D.; and Feuchtgruber, H., "Titan's Atmosphere from ISO Mid-Infrared Spectroscopy", *Icarus, vol. 161*, pp. 383-403, 2003.

Hourdin, F.; Talagrand, O.; Sadourny, R.; Courtin, R.; Gautier, D.; and McKay, C.P., "Numerical Simulation of the General Circulation of the Atmosphere of Titan", *Icarus, vol. 117*, pp. 358-374, 1995.

Justus, C. G.; Duvall, A.; Keller, V. W., "Engineering-Level Model Atmospheres for Titan and Mars", Proceedings of the International Workshop on Planetary Probe Atmospheric Entry and Descent Trajectory Analysis and Science, Lisbon, Portugal, 6-9 October, 2003, ESA SP-544, February, 2004.

NASA Engineering and Safety Center Technical Assessment Report	Document #: RP-05-67	Version: 1.0
Title: **Independent Technical Assessment of Cassini/Huygens Probe Entry, Descent and Landing (EDL) at Titan**	Page #: 54 of 116	

Strobel, D. F. and B. Sicardy, "Gravity Wave and Wind Shear Models", in *Huygens, Science, Payload and Mission*, ESA SP-1177, August, 1997.

Yelle, R. V.; D. F. Strobel; E. Lellouch; and D. Gautier, "Engineering Models for Titan's Atmosphere", in *Huygens, Science, Payload and Mission*, ESA SP-1177, August, 1997.

Yelle, R. V., "Workshop Review and Atmosphere Model Presentation", Titan Atmospheric Model T0 Workshop, NASA Goddard Space Flight Center, Greenbelt, MD, September 8-9, 2004.

Yelle, R. V., "General Discussion, Plans for TA Model Atmospheres", Titan Atmospheric Workshop - TA Results, Monrovia, CA, November 15, 2004.

Appendix D

Attachment A

Trip Report

Titan Atmospheric Model T0 Workshop
September 8-9, 2004
NASA Goddard Space Flight Center
Greenbelt, MD

C. G. Justus
ED44/Morgan Research
NASA Marshall Space Flight Center

Wednesday, September 8

Overview Presentation by Jean-Pierre Lebreton

Huygens design uses the Huygens Recovery Task Force (HRTF or Flasar) wind model, with peak of 100-150 m/s at ~ 220 km. Although consensus is now that winds are prograde, the original system design used both prograde and retrograde winds. Christophe Lowe of Ecole Centrale de Paris is leading heat flux reassessment (for fluxes up to 2000 kW/m²); final results are expected by end of September. Preliminary tests of UV heat flux (CN radiation) at ARC look promising. Higher flux rates will be tested soon. Roger Yelle is to lead the Titan Atmosphere Model Working Group (TAMWG), to produce an initial T0 model by 9/30/04, with updates 11/23/04, 12/17/04, and 1/3/05. Lebreton mentioned a new "Independent NASA/ESA Review" being conducted, with final results expected by 11/11/04. Final go/no-go decision is expected on 12/2/04. If no-go for separation on 12/23, there are other opportunities about one month and five months later.

CIRS Presentation by Barney Conrath and Rich Achterburg

The Composite Infrared Spectrometer (CIRS) measurements during the "T0" encounter (July 4, 2004) measured temperature versus pressure, with T ~ 182 K at 0.3 mb, T ~ 160 K at 2 mb, and T ~ 140 K at 10 mb (~ 100 km altitude). Polar temperatures at 1 mb are ~ 165 K, while equatorial temperatures are ~ 170 K at this level. A large, very deep, warm patch was observed

near the equator and 0° longitude (just East of the sub-Saturn point). Observed characteristics of this feature are difficult to explain dynamically.

Occultation Presentation by Bruno Sicardy

Stellar occultation results from November 2003 gave densities slightly higher than mean Yelle values in the 300-500 km height range, but well within the Yelle maximum envelope. Winds inferred from central flash observations were ~220 m/s at 0.25 mb, with a "northern jet" near 50 N. No evidence for a southern jet was observed. Winds were inferred to approach zero near both poles. Dust optical depths were inferred to have a wavelength exponent of 1.6±0.2, which implies large particles (difficult to explain at high altitudes, since they should settle out quickly).

Cloud Motion Winds Presentation by Tony DelGenio

The Imaging Science Subsystem (ISS) measured winds of 34 m/s prograde and 9 m/s poleward from a large patch of convective clouds near the south pole. These winds are likely in the troposphere, and may be at the tropopause (anvil clouds at convective tops).

Meeting Jim Frautnick and Denis Bogan

During the break, Jim Frautnick (JPL) and Denis Bogan (NASA HQ, Cassini Program Office) introduced themselves. Frautnick says Gentry Lee is leading the JPL re-analysis team. Bogan indicated that he had reviewed our Huygens re-analysis proposal statement of work. Frautnick was very interested in making sure that we would be exchanging data and models with Roger Yelle's TAMWG.

Heterodyne Winds Presentation by Theodor Kostuik

Kostuik reported results of very high resolution infrared spectroscopy (heterodyne) measurement of Titan winds, as large as 170 ± 90 m/s.

Spacecraft Torque Presentation by Allan Lee

Lee (JPL) discussed plans to try to measure atmospheric density from accelerometer-derived spacecraft torques, during the "TA" and "T1" flyby at 1200 km altitude. Plans for subsequent flyby altitudes are 1000 km for "T3" and 950 km for "T5". These altitudes may be adjusted, depending on results from the TA and T1 passes.

Thursday, September 9

Parachute System Review by Lebreton

Because of the failure of the Genesis parachute system, Lebreton gave an extensive review of the Huygens parachute-deploy system and battery system. There are three Huygens chutes - a pilot chute, the main chute, and a stabilizer chute (to keep probe descent stable, but at a faster rate than allowed by the main chute, so the probe can reach the surface within a reasonable elapsed time). Primary chute deploy sequences are controlled by software enabling, from accelerometer measurements. A hardware (g-switch) backup system manages the chute deploy sequence if there is a failure of the software system. A system time-out mechanism acts as a secondary backup if both software and hardware systems should fail. There are five batteries. One is being activated next week as a test. The complete mission can be done with only four working batteries. If there are only three working batteries, the mission can still be accomplished, but probe switch-on and warm-up would be delayed in that case.

Workshop Review and Atmosphere Model Presentation by Roger Yelle

Workshop consensus was that methane is saturated in the troposphere, with stratospheric methane mole fraction of 1 to 3%, and argon mole fraction of 0 to 7%. Yelle's preliminary atmosphere model, incorporating latest CIRS and occultation measurements, is slightly higher than original Yelle mean atmosphere, but well within Yelle maximum profile. With CIRS and occultation winds of ~ 140 m/s and heterodyne winds of 170 ± 90 m/s, winds may be larger than have been assumed in the HRTF wind model. No evidence has been found for large waves near 200 km altitude, so the Strobel and Sicardy model continues to be conservative. Titan Global Reference Atmosphere Model results (shown by Lebreton) are consistent with the Strobel and Sicardy model, and perhaps a little more "benign", especially with regard to parachute deploy parameters.

Follow-on Workshop Plans Presentation by Scott Bolton

Two more flybys (called TA and TB) are scheduled before Huygens release. TA will be at an altitude of 1200 km on day 300 of 2004. TB altitude is 2400 km. It is not clear if TB data (taken on day 347 of 2004) can be turned around in time to influence the Huygens go/no-go decision. A tentative TA workshop is planned for 11/15/04 at JPL. After Huygens nominal release time, there may be a TB workshop in late January or February of 2005. A first Huygens Science Workshop may also be held in early spring of 2005 (possibly in conjunction with the European Geoscience Union meeting).

Effects of Iaepetus Mass Uncertainty Presentation by Dennis Matson

The initial pass by Iaepetus indicated it may have a mass significantly outside previous error estimates. This mass uncertainty could affect the Huygens trajectory after release on the original schedule. A distant (10^6 km) Iaepetus flyby may provide a mass update. If an apparent problem persists, Huygens separation could be postponed, allowing another Iaepetus flyby on 1/1/05 to provide additional data. Huygens release could be postponed as much as about five months.

Plans for Follow-Up TAMWG Telecom

After the workshop was over, a brief meeting was held with Roger Yelle, Jim Frautnick, and Jean-Pierre Lebreton. Yelle will provide atmospheric data to us as soon as it is available (either by e-mail or through the Cassini password-controlled web site, if we can arrange for access to that site). Lebreton asked if we could support a TAMWG telecom that will be held sometime the week of 9/20.

Appendix D

Attachment B

Trip Report
Titan Atmospheric Workshop - TA Results
Monrovia, CA
November 15, 2004

C. G. Justus
Morgan Research
NASA Marshall Space Flight Center

Introduction and Huygens Status - Jean Pierre LeBreton

New recommended nominal atmospheric mole fractions are Ar=0.0%, CH_4 about 2%. Wind gusts were suggested as a possible problem for Huygens parachute deploy but it was not made clear why wind gusts, and not just density gradients, were of concern for this issue. ESA/JPL Huygens independent review report is due 22 November with a presentation to JPL Director, Dr. Elachi scheduled for 29 November. LeBreton expressed appreciation to MSFC In-Space Propulsion Program for prior support of the Titan Systems Analysis Team, and especially support of the aerothermal analysis teams at NASA ARC and NASA Langley and the trajectory and guidance team at NASA Langley. Because of that support, these teams have well-developed and validated tools in place that are proving extremely valuable in the Huygens review activities.

Current Status of Model Atmospheres (T0) - Roger Yelle

Yelle reviewed the status of latest (T0 Titan pass results) model atmospheres. TA pass atmosphere model updates will be made available in late November. Subsequent atmosphere model updates will also be made available in both December and January.

Imaging Science Subsystem (ISS) - Andy Ingersoll

ISS reported a prograde (Eastward) mean wind of 34 m/s ($\pm$ 10 m/s) in the upper troposphere ($\sim$ 40 km altitude). Surface winds were inferred to be a few (up to $\sim$ 10) m/s, based on images of surface wind streaks.

	NASA Engineering and Safety Center Technical Assessment Report	Document #: RP-05-67	Version: 1.0
Title: **Independent Technical Assessment of Cassini/Huygens Probe Entry, Descent and Landing (EDL) at Titan**			Page #: 60 of 116

Composite Infra-Red Spectrometer (CIRS) - Mike Flasar

Due to an instrument software problem, much less CIRS data was obtained from the TA pass than the T0 pass. Because of the closer approach at TA, available TA data are at significantly better spatial resolution. T0 temperature sounding data, when used in thermal wind relations, imply prograde winds slightly in excess of 100 m/s near 200 km altitude. Winds in the northern hemisphere (well away from the Huygens landing site) may be as large as 200 m/s at these altitudes. TA data were not of sufficient quantity to derive winds, but lat-lon variations of temperature can be inferred. TA temperatures are slightly warmer than T0 results. When reduced by 8%, because of recent updates in CH_4 absorption line strengths, T0 results for CH_4 mole fraction were 1.7% ± 0.5%. TA results are 1.75% ± 0.75%.

Radar - Ralph Lorenz

On the TA pass, the radar measured surface temperature 93K ± 5K and N_2 optical depth = 0.03 ± 0.01. This optical depth is proportional to surface pressure, but the error bound is not adequate to constrain current surface pressure versus pressure from earlier Voyager flyby results.

Visual and Infrared Mapping Spectrometer (VIMS) - Pierre Drossart

CH_4 measurements were found to be consistent with 2% (± 1%) mole fraction, and consistent with concentrations having uniform distribution with altitude and latitude.

Ion and Neutral Mass Spectrometer (INMS) - Hunter Waite

Composition and density measurements, down to 1174 km periapsis, were projected downward to 950 km (height of some subsequent Titan passes). Density at 1174 km was measured as 4.6×10^{-11} kg/m^3; density at 950 km was inferred to be 1.4×10^{-9} kg/m^3. Ar mole fraction was observed as < 10 ppm (< 0.001%), with CH_4 mole fraction = 2.0% ± 0.2%. An isothermal temperature of 149K was observed between 950 and 1200 km. Exobase height was observed to be 1400 km, with homopause height = 1200 km. There is some evidence of diffusive separation of CH_4 above 1200 km. Waves were observed, with amplitude ~ 10 K (7%) and vertical wavelength ~ 200 km. Observations imply a mass density slightly less than inferred from Vervack et al. Icarus paper.

Texas Echelon Cross Echelle Spectrograph (TEXES) - Caitlin Griffith

Observed spectra were fit with CH_4 mole fractions ranging from 1% to 3%. Best fit seems to be ~ 2%, but they have not yet tried values between 2 and 3%. Temperature was inferred to be 140-157 K near 600 km altitude.

	NASA Engineering and Safety Center Technical Assessment Report	Document #: RP-05-67	Version: 1.0
Title: **Independent Technical Assessment of Cassini/Huygens Probe Entry, Descent and Landing (EDL) at Titan**			Page #: 61 of 116

Ultra-Violet Imaging Spectrograph (UVIS) - Don Shemansky

UVIS sees some evidence of latitude variation in CH_4 (T0 gave an equatorial pass; TA gave a pole-to-pole pass). Best estimate for CH_4 mole fraction is 3.2%, but it could be as low as 2%, when errors are considered.

Spacecraft Torque and Density Results - Allan Lee

Lee used torques, derived from thruster firings during TA pass, to infer density down to 1174 km. These values were projected downward to 950 km (for interest on future Titan passes). Density at 1174 km was estimated to be 2.0×10^{-10} kg/m^3 (a factor of 4 larger than observed by INMS). Downward projection to 950 km yielded 2.7×10^{-9} kg/m^3 (a factor of 2 larger than inferred from INMS). Follow-up analysis is required to resolve this discrepancy between the torque and INMS techniques, since the torque-inferred density at 950 km exceeds the critical density that would induce spacecraft tumble. If this density value holds up, then the 950 km planned periapsis on future Titan passes would have to be raised.

General Discussion, Plans for TA Model Atmospheres - Roger Yelle

Final recommendations of the Titan Atmosphere Model Working Group (TAMWG) team were: Use Ar mole fraction = 0.0%. Use CH_4 mole fraction minimum = 1.2%, maximum = 2.3% (uniform distribution). The TAMWG will generate TA profiles (by about end of November), using estimated temperature uncertainty, to derive not just mean density profile, but also min and max density profiles for given CH_4 concentration. It was not clear whether the final product would be three profiles (min/avg/max density at average CH_4) or nine profiles (min/avg/max density at min/avg/max CH_4 concentration). The group's estimate of mean wind above 200 km was 130 m/s, with standard deviation (perturbation magnitude) of 30 m/s. With ISS estimates of 34 m/s at 40 km, and < 10 m/s at the surface, this made the group recommendations very close to Titan-GRAM wind profiles of mean and standard deviation. The final recommendation of the TAMWG was to adopt Titan-GRAM wind model for Huygens entry studies. Titan-GRAM will continue to be used for trajectory and parachute deploy analyses, and to provide environments estimates to the thermal analysis teams.

Appendix E. POST2-Based Flight Simulation

Prepared by
Scott A. Striepe and Jody L. Fisher
NASA LaRC

E.0 Introduction

A six degree-of-freedom (6DOF) atmospheric entry and three degree-of-freedom (3DOF) parachute descent trajectory of the Huygens probe was simulated in Program to Optimize Simulated Trajectories II (POST2) as part of the NESC independent technical assessment. The POST2-based flight simulation incorporated a 6DOF aerodynamics model, a Titan-GRAM atmospheric model, and a parachute model. These models are discussed in more detail in the following sections of this Appendix:

E.1   Core Trajectory Simulation in POST2 - discusses flight and systems studies heritage description and a detailed phase layout of the nominal 6DOF entry and 3DOF parachute descent trajectory.

E.2   Initial States for Entry Vehicle - discusses the Huygens probe in conjunction with a tabulated list of simulation inputs.

E.3   Entry Vehicle Aerodynamics Database Subroutine – discusses the 6DOF aerodynamic database developed in support of the Huygens trajectory simulation.

E.4   Parachute Model Characteristics

E.5   Titan-GRAM Atmospheric Model – discusses implementation, input list, and density and wind dispersions from the nominal POST2 trajectory.

E.6   Huygens Probe Mass Property Data

E.7   Aeroheating Parameter Calculations - discusses the aeroheating parameters calculated in the POST2 trajectory.

NASA Engineering and Safety Center Technical Assessment Report	Document #: RP-05-67	Version: 1.0
Title: **Independent Technical Assessment of Cassini/Huygens Probe Entry, Descent and Landing (EDL) at Titan**	Page #: 63 of 116	

E.1 Core Trajectory Simulation in POST2

E.1.1 Flight and Systems Studies Heritage of POST2

The Program to Optimize Simulated Trajectories II (POST2) is a generalized point mass, discrete parameter targeting and optimization trajectory program. POST2 has the ability to simulate three and six degree-of-freedom (3DOF and 6DOF) trajectories for multiple vehicles in various flight regimes. POST2 also has the capability to include different atmosphere, aerodynamics, gravity, propulsion, parachute and navigation system models. Many of these models have been used to simulate the entry trajectories for previous missions (i.e., MER, Genesis, Mars Pathfinder) as well as current and planned NASA missions (Stardust, Mars Phoenix, and Mars Science Laboratory). A variety of system studies have been conducted and their atmospheric trajectories simulated in POST2 including aerocapture at Titan, Neptune and Venus. In addition to the systems studies conducted, an initial simulation of the Huygens probe entry trajectory was generated for an ongoing graduate research program, which provided the groundwork for the Huygens probe flight simulations used to support the NESC ITA. This simulation was used to produce single trajectory data and was an integral element of the Monte Carlo analyses (discussed in Appendix F).

E.1.2 6DOF Entry-3DOF Parachute Descent Trajectory

POST2 was used to simulate the Huygens entry trajectory into Titan. The simulation included vehicle geometric parameters, Titan's gravity and atmosphere models, attitude inputs and initial states (refer to Table E-1). The Titan-GRAM atmosphere model was initialized at the atmospheric interface event; Table E-2 shows the Titan-GRAM inputs used in the simulation. Parachute trigger criteria and inflation models for the pilot, main, and drogue parachutes were also included in the simulation.

The main sequential phases in the simulation were as follows:

- initialization
- atmospheric interface
- hypersonic entry
- pilot parachute deployment, inflation, and flight
- main parachute deployment, inflation, and flight
- heat shield release
- main parachute release
- stabilizer drogue parachute deployment, inflation, and flight
- surface impact

Figure E-1 is a plot of the 6DOF-3DOF descent profile for the nominal trajectory simulated in POST2. Figure E-2 contains plots of the entry characteristics (total angle-of-attack, angular rates, dynamic pressure and decelerations) for the nominal 6DOF-3DOF trajectory.

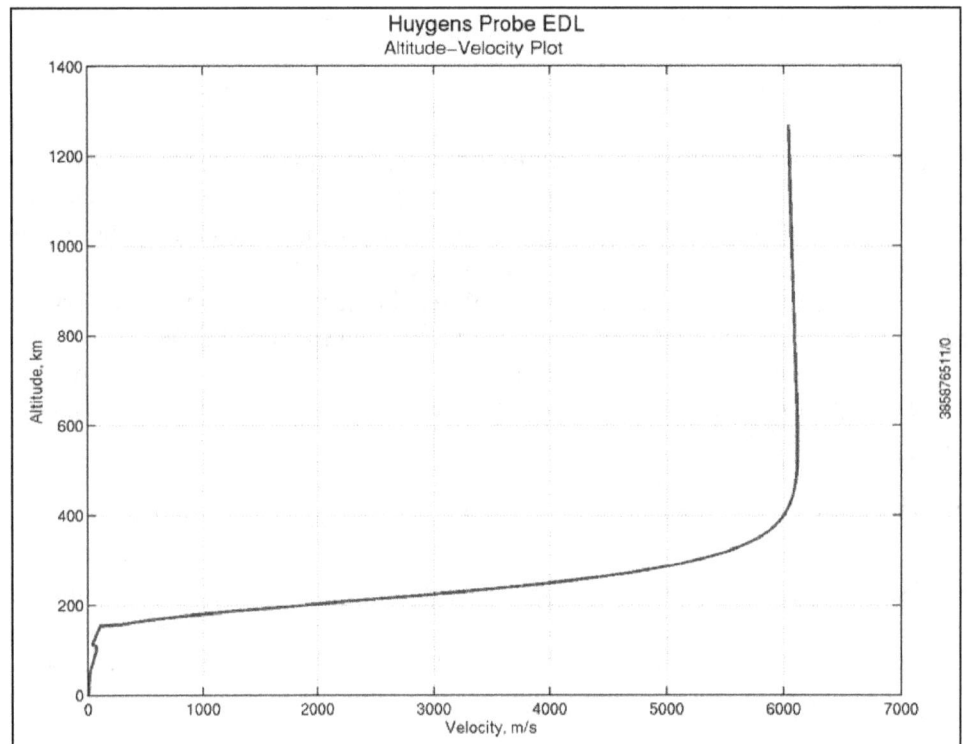

Figure E-1. Nominal 6DOF-3DOF Descent Profile

NASA Engineering and Safety Center Technical Assessment Report	Document #: RP-05-67	Version: 1.0
Title: Independent Technical Assessment of Cassini/Huygens Probe Entry, Descent and Landing (EDL) at Titan		Page #: 65 of 116

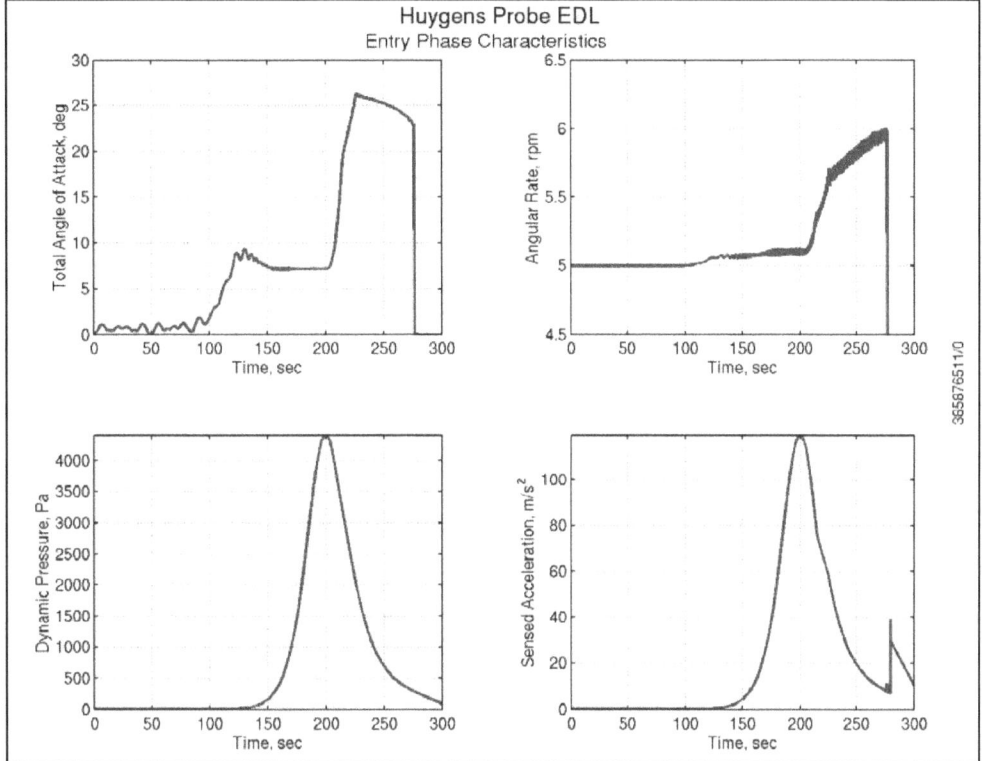

Figure E-2. Nominal 6DOF-3DOF Entry Phase Characteristics

E.2 Initial States for Entry Vehicle

The nominal states and the corresponding covariances for the entry vehicle were provided by JPL. The position and velocity coordinates were provided in the Titan Equatorial, Prime Meridian of the Epoch frame. The coordinate frame was established at 9 hours, 6 minutes, and 0 seconds on 14 January 2005. Table E-1 shows the nominal initial state used in the simulation.

NASA Engineering and Safety Center
Technical Assessment Report

Document #:
RP-05-67

Version:
1.0

Title:
Independent Technical Assessment of Cassini/Huygens Probe Entry, Descent and Landing (EDL) at Titan

Page #:
66 of 116

Table E-1. POST2 Simulation Inputs

Input Parameter	Value
Initial Position Vector Components (m)	x(1) = -3.302882651e+06 x(2) = 2.466321784e+06 x(3) = -6.614883268e+05
Initial Velocity Vector Components (m/sec)	vx(1) = 5.792965058e+03 vx(2) = -1.434607255e+03 vx(3) = 5.991841457e+02
Capsule Mass (kg)	320
Probe Reference Length (m)	2.7
Probe Reference Area (m^2)	5.726
CG Location Components* (m)	x(axial) = 0.47176 y(lateral) = 0.001539 z(lateral) = 0.004914
Moments of Inertia Components** (kg-m^2)	Ixx(roll) = 127.97 Iyy(pitch) = 75.85 Izz(yaw) = 71.9
Products of Inertia Components** (kg-m^2)	Ixy = 0.45 Ixz = 0.096 Iyz = -0.338
Initial Attitude** (deg)	Roll = 80.36 Pitch = 18.42 Yaw = 7.46
Initial Attitude Rates** (deg/sec)	Roll Body Rate = 43.725 Yaw Body Rate = 0.0 Pitch Body Rate = 0.0
Titan Equatorial Radius (m)	2575.e3
Titan Polar Radius (m)	2575.e3
Titan Gravitational Constant (m^3/sec^2)	8.977947e+12
Titan Rotation Rate (rad/sec)	4.560678e-06

* origin at vehicle nose in body reference frame
** about the body axes centered on the vehicle CG

E.3 Entry Vehicle Aerodynamics Database Subroutine

NASA developed the Huygens probe 6DOF aerodynamic database based on Genesis data and Huygens probe-shape ballistic range data (refer to Appendix B). The database was then incorporated into an aerodynamics subroutine and used in the 6DOF-3DOF trajectory simulation. The ESA-generated aerodynamics was also included as a separate subroutine used in the simulation.

E.4   Parachute Model Characteristics

The descent system is comprised of three Disk-Gap-Band parachutes: the pilot, main, and stabilizing drogue. Appendix C discusses the drag model sets and opening loads of the parachute system in greater detail. The primary and backup trigger for the pilot parachute deployment was based on an acceleration logic provided by ESA and was included in the POST2 simulation. Parachute inflation models were incorporated into the POST2 simulation. The sequence for the pilot parachute modeled in POST2 started with acceleration logic for deployment initiation, mortar fire, inflation model, and fully inflated flight. The main and stabilizer drogue parachutes had similar sequences except that the deployment trigger was based on a timer and not on acceleration.

E.5   Titan-GRAM Atmospheric Model

Version 1.0 of the Titan-GRAM atmospheric model was implemented into POST2, with updates from Cassini measurements (the T0 and TA atmospheric profiles). Appendix D provides the nominal and dispersion inputs for the trajectory simulation, including the T0 and TA average, high and low inputs (Table E-2 shows the Titan-GRAM inputs used in the simulation). More information about Titan atmospheric properties and Titan-GRAM input descriptions are included in Appendix D.

The amount of atmospheric methane (CH_4) used in the simulation was kept at a molar fraction of 3% of the Titan atmosphere. Figures E-3 through E-6 show percent variation in the density and wind (East-West and North-South) perturbations using the GCM and TA profile data in Titan-GRAM with respect to the nominal 6DOF-3DOF trajectory (red curve on Figures E-5 and E-6).

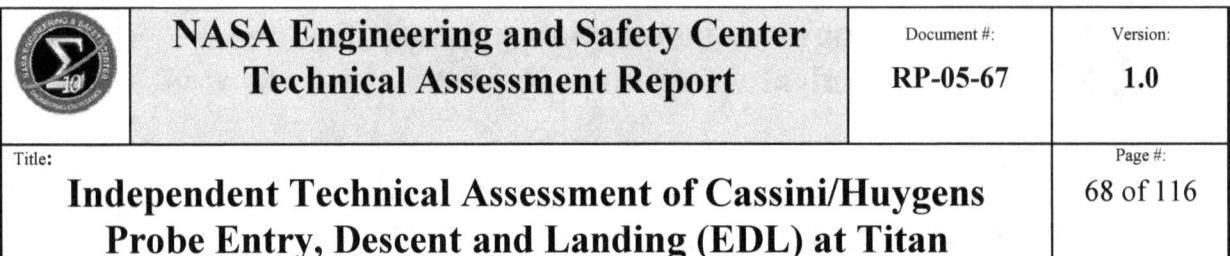
Table E-2. Titan-GRAM Inputs

Titan-GRAM Input	Value
IERT	0
IUTC	1
MONTH	1
MDAY	14
MYEAR	2005
IHR	9
IMIN	6
SEC	0.29
NPOS	100
LonEast	1
Fminmax	0.0
IFMM	1
NR1	1001
rpscale	1.0
NMONTE	1
iup	0
corlmin	0.0
fmolmeth	3.0
DELHGT	0

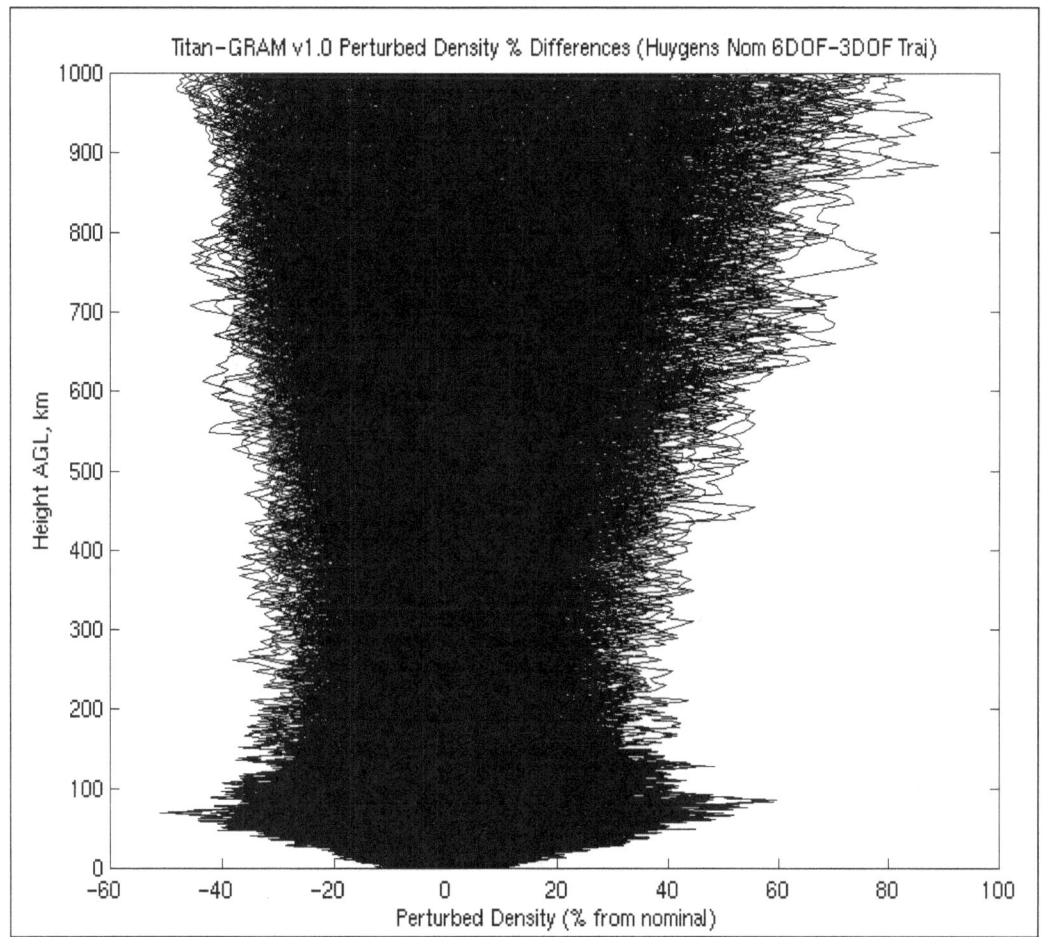

Figure E-3. GCM Perturbed Density % from Nominal

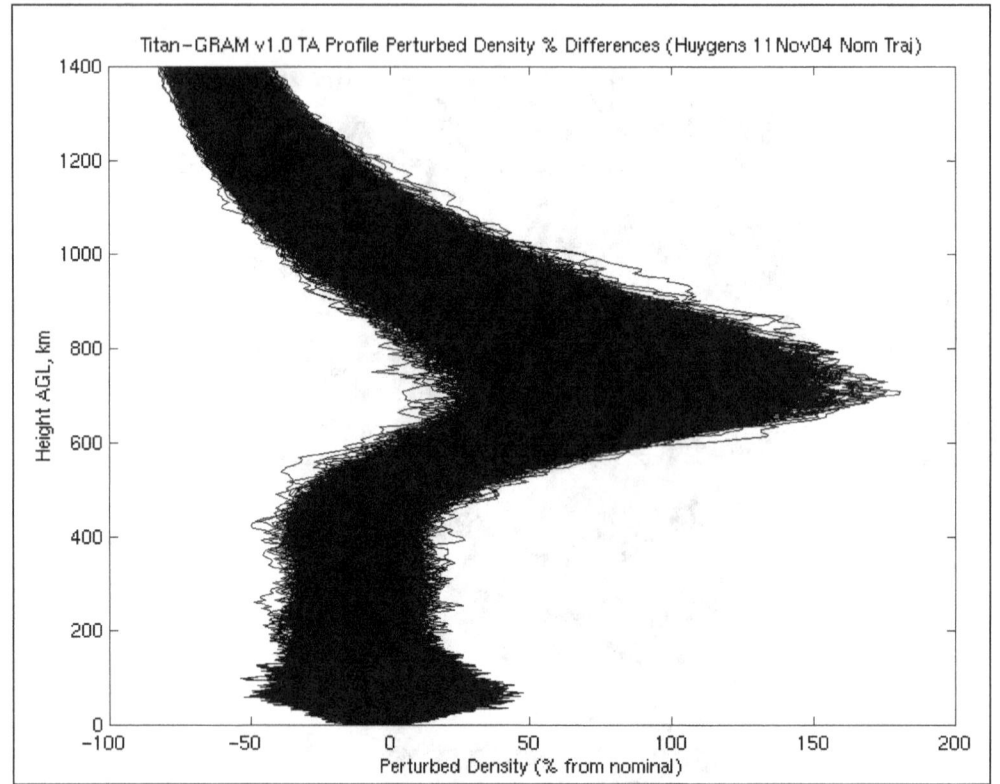

Figure E-4. TA Profile Perturbed Density % from Nominal

NASA Engineering and Safety Center Technical Assessment Report	Document #: RP-05-67	Version: 1.0
Title: Independent Technical Assessment of Cassini/Huygens Probe Entry, Descent and Landing (EDL) at Titan		Page #: 71 of 116

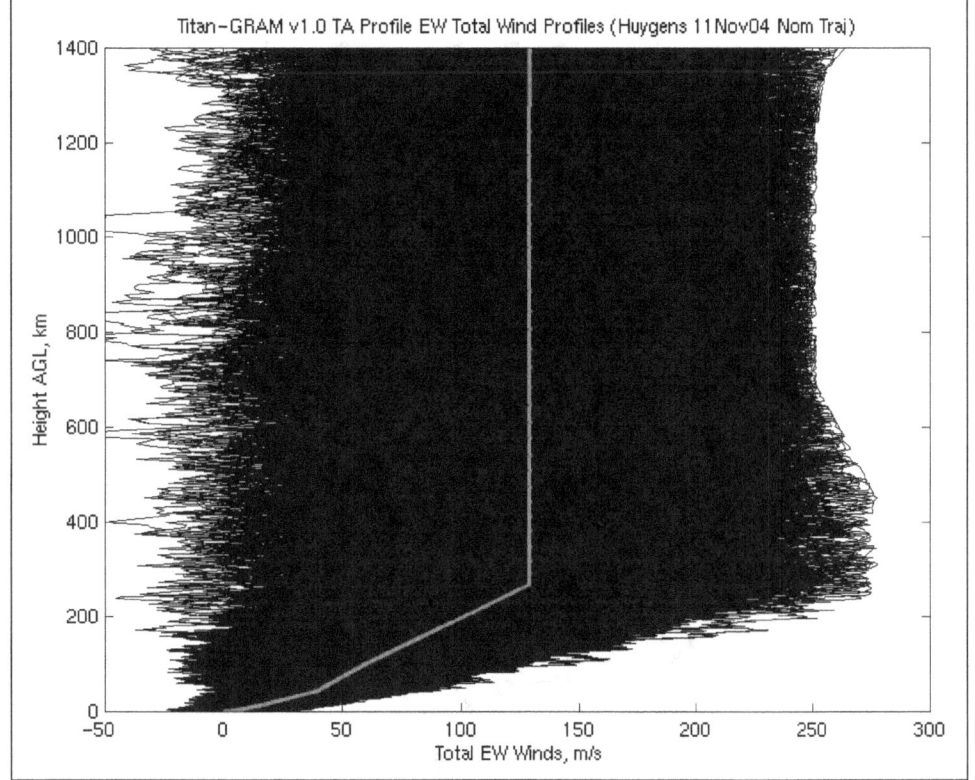

Figure E-5. TA Profile Perturbed EW Winds from Nominal

NASA Engineering and Safety Center Technical Assessment Report	Document #: RP-05-67	Version: 1.0
Title: **Independent Technical Assessment of Cassini/Huygens Probe Entry, Descent and Landing (EDL) at Titan**		Page #: 72 of 116

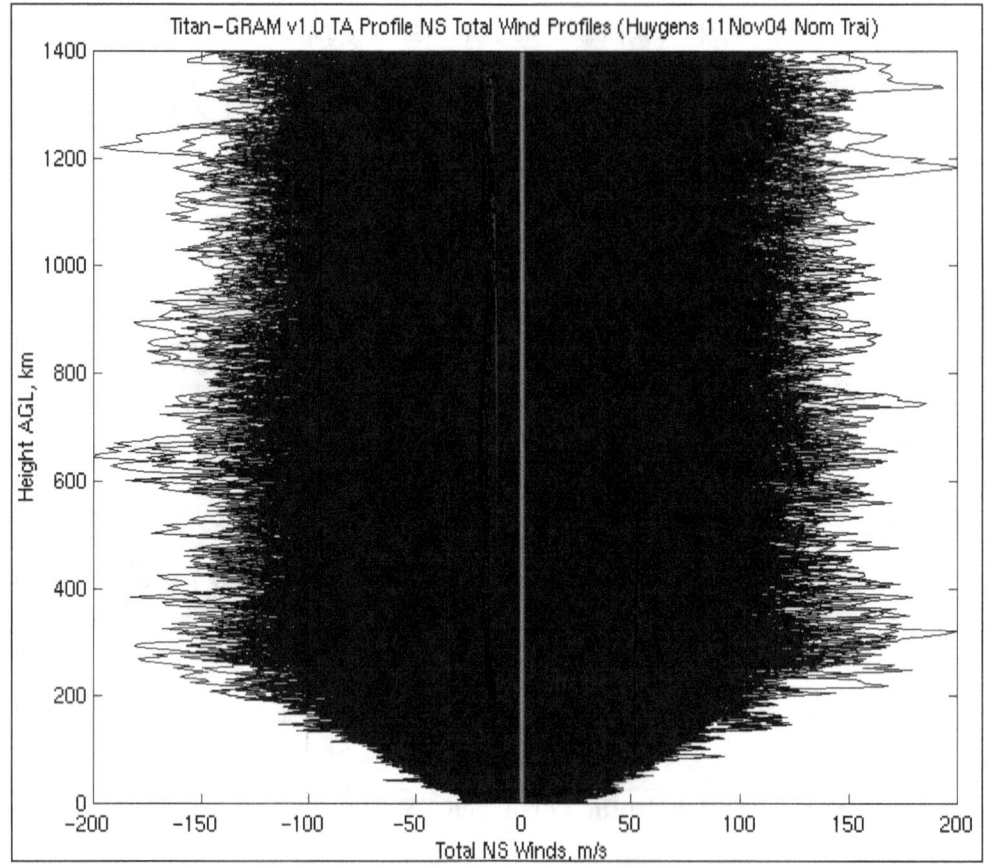

Figure E-6. TA Profile Perturbed NS Winds from Nominal

E.6 Huygens Probe Mass Property Data

The mean mass property values for the entry vehicle were obtained from ESA documents and input in the simulation (refer to Table E-1). The entry mass used was 320 kg, with the appropriate center of gravity (cg) location (axial and lateral), inertias (roll, pitch and yaw), and cross-product inertias.

E.7 Aeroheating Parameter Calculations

The aeroheating parameters were divided into three separate, simplified calculations in the POST2 simulation: radiative, laminar and turbulent. Time-history curve-fits from detailed calculations at various points along representative trajectories were generated for use in POST2. These curve-fits are suitable for initial assessment, trends and isolation of cases for detailed analysis. More information on the probe aeroheating is given in Appendix G.

NASA Engineering and Safety Center Technical Assessment Report	Document #: RP-05-67	Version: 1.0
Title: **Independent Technical Assessment of Cassini/Huygens Probe Entry, Descent and Landing (EDL) at Titan**	Page #: 73 of 116	

Appendix F.   Monte Carlo Analyses and Results

Prepared by
Scott A. Striepe and Richard W. Powell
NASA LaRC

F.0   Introduction

The simulation described in <u>Appendix E</u> was used in a Monte Carlo analysis of the Huygens-probe entry, descent, and impact on Titan. The analyses and results are discussed in more detail in the following sections of this Appendix:

F.1   Monte Carlo Analysis

F.2   Monte Carlo Dispersed Inputs

F.3   Monte Carlo Results

F.1   Monte Carlo Analysis

The Monte Carlo technique involves the variation of key input parameters to encompass the level of uncertainty in these inputs. That is, once the range of uncertainty in the inputs was established, random numbers were used to determine the specific input value selected for a given simulation run. Several thousand runs were made in this fashion and statistics of the resulting outputs were analyzed.

F.2   Monte Carlo Dispersed Inputs

The Monte Carlo dispersed inputs are assumed to have a certain distribution (e.g., Gaussian, Uniform, etc.) with a given mean and extreme values. The discipline experts on the NESC team were consulted to define ranges for the various input variables. Other inputs were taken from previously defined ranges. For example, the dispersions of the mass properties were developed from previous NASA missions, such as Genesis and MER, due to the unavailability of those dispersions from the ESA. Table F-1 shows the inputs and dispersion ranges used in the Monte Carlo assessments.

Table F-1. Huygens Titan Probe 6DOF Entry Dispersions

Quantity	Nominal Value	Distribution Type	3-σ or min/max
Mission Uncertainty			
Initial Roll Angle, deg	80.36	Gaussian	2.7
Initial Pitch Angle, deg	18.42	Gaussian	2.7
Initial Yaw Angle, deg	7.46	Gaussian	2.7
Initial Roll Rate, deg/sec	43.725	Gaussian	10%
Initial Pitch Rate, deg/sec	0.0	Gaussian	0.4
Initial Yaw Rate, deg/sec	0.0	Gaussian	0.4
Aerodynamic Uncertainty			
Probe Axial Force Coeff Mult. ($K_n \geq 0.1$)	1.0	Gaussian	5 %
Probe Normal Force Coeff Incr ($K_n \geq 0.1$)	0	Gaussian	0.01
Probe Axial Force Coeff Mult. (Mach > 10)	1.0	Gaussian	3 %
Probe Normal Force Coeff Incr (Mach > 10)	0	Gaussian	0.01
Probe Axial Force Coeff Mult. (Mach < 5)	1.0	Gaussian	10 %
Probe Normal Force Coeff Incr (Mach < 5)	0	Gaussian	0.01
Probe Pitch Moment Coeff Incr. ($K_n \geq 0.1$)	0	Gaussian	0.005
Probe Pitch Moment Coeff Incr. (Mach > 10)	0	Gaussian	0.003
Probe Pitch Moment Coeff Incr. (Mach < 5)	0	Gaussian	0.005
Probe Pitch Damping Coeff Incr. (Mach > 6)	0	Gaussian	0.15
Probe Pitch Damping Coeff Incr. (Mach < 3)	0	Gaussian	0.15
Pilot Parachute Drag Coeff Mult.	1.0	Gaussian	0.05
Main Parachute Drag Coeff Mult.	1.0	Triangular	0.1
Drogue Parachute Drag Coeff Mult.	1.0	Triangular	0.1
Parachute Opening Load Factor	1.42	Uniform	0.05
Mass Property Uncertainty			
Mass, kg	320.0	Gaussian	1.0
Axial CG position, m	0.47176	Uniform	0.03175
Lateral CG position (Y), m	0.00154	Uniform	0.0069
Lateral CG position (Z), m	0.00491	Uniform	0.0069
Ixx, kg-m^2	127.97	Gaussian	10 %
Iyy, kg-m^2	75.85	Gaussian	10 %
Izz, kg-m^2	71.9	Gaussian	10 %
Ixy, kg-m^2	0.45	Gaussian	2.0
Ixz, kg-m^2	0.096	Gaussian	2.0
Iyz, kg-m^2	-0.338	Gaussian	2.0
Atmospheric Uncertainty			
Initial Seed Value	1	Uniform	1/29999
Fminmax input	0	Uniform	+/- 1.0
Parachute Deploy Device Uncertainty			
Accelerometer #1 Constant Mult.	1.0	Uniform	5.37%
Accelerometer #2 Constant Mult.	1.0	Uniform	5.37%
Accelerometer #3 Constant Mult.	1.0	Uniform	5.37%
Accelerometer #1 Analog Constant Mult.	1.0	Uniform	4.68%
Accelerometer #3 Analog Constant Mult.	1.0	Uniform	4.68%

F.3   Monte Carlo Results

Many Monte Carlo runs were completed in support of this activity. Table F-2 summarizes all of the cases considered. Some of these cases were used to support the primary findings from this team as indicated in the sections below.

Table F-2. Monte Carlo Cases Completed

Parachute Model	Aerodynamics Model	Mean Entry FPA,deg	Entry FPA Variation, deg	Titan Atm Profile Used	PDD Used
ESA	NASA	-65.15	3.0	Titan-GRAM	NASA
ESA	NASA	-63.0	0.0	Titan-GRAM	NASA
ESA	NASA	-65.15	0.0	Titan-GRAM	NASA
ESA	NASA	-67.0	0.0	Titan-GRAM	NASA
ESA	NASA	-65.15	1.5	Titan-GRAM	NASA
ESA	NASA	-66.5	1.5	Titan-GRAM	NASA
ESA	NASA	-63.5	1.5	Titan-GRAM	NASA
ESA	ESA	-65.15	3.0	Titan-GRAM	NASA
NASA 1	NASA	-65.15	3.0	Titan-GRAM	NASA
NASA 1	NASA	-65.15	1.5	Titan-GRAM	NASA
NASA 1	NASA	-66.5	1.5	Titan-GRAM	NASA
NASA 1	NASA	-63.5	1.5	Titan-GRAM	NASA
NASA 1	NASA	-61.5	1.5	Titan-GRAM	NASA
NASA 1	NASA	-59.5	3.0	Titan-GRAM	NASA
NASA 1	NASA	-55.4	1.5	Titan-GRAM	NASA
NASA 1	ESA, NASA uncerts	-65.15	3.0	Titan-GRAM	NASA
NASA 1	NASA	-65.15	3.0	T0 Average	NASA
NASA 1	NASA	-65.15	3.0	T0 High	NASA
NASA 1	NASA	-65.15	3.0	T0 Low	NASA
NASA 1	NASA	-65.15	3.0	T0 Average	ESA
NASA 1	NASA	-65.15	3.0	T0 High	ESA
NASA 1	NASA	-65.15	3.0	T0 Low	ESA
Vorticity 1	NASA	-65.15	3.0	Titan-GRAM	ESA
Vorticity 1	NASA	-65.15	3.0	T0 Average	ESA
NASA 2	NASA	-65.15	3.0	Titan-GRAM	ESA
NASA 2	NASA	-65.15	3.0	T0 Average	ESA
NASA 2	NASA	-65.15	3.0	T0 Low	ESA
NASA 2	NASA	-65.15	3.0	T0 High	ESA

F.3.1 Parachute Opening Loads

While analyses using the NASA 1 parachute model set predict higher drogue and main parachute peak opening loads than those using the ESA model set, they are still within the design specifications of the system. Figure F-1 shows the main parachute opening loads when using the ESA parachute model set. In contrast, Figure F-2 shows loads from the NASA parachute model set. Note that both loads are below the maximum limit of 17600 N.

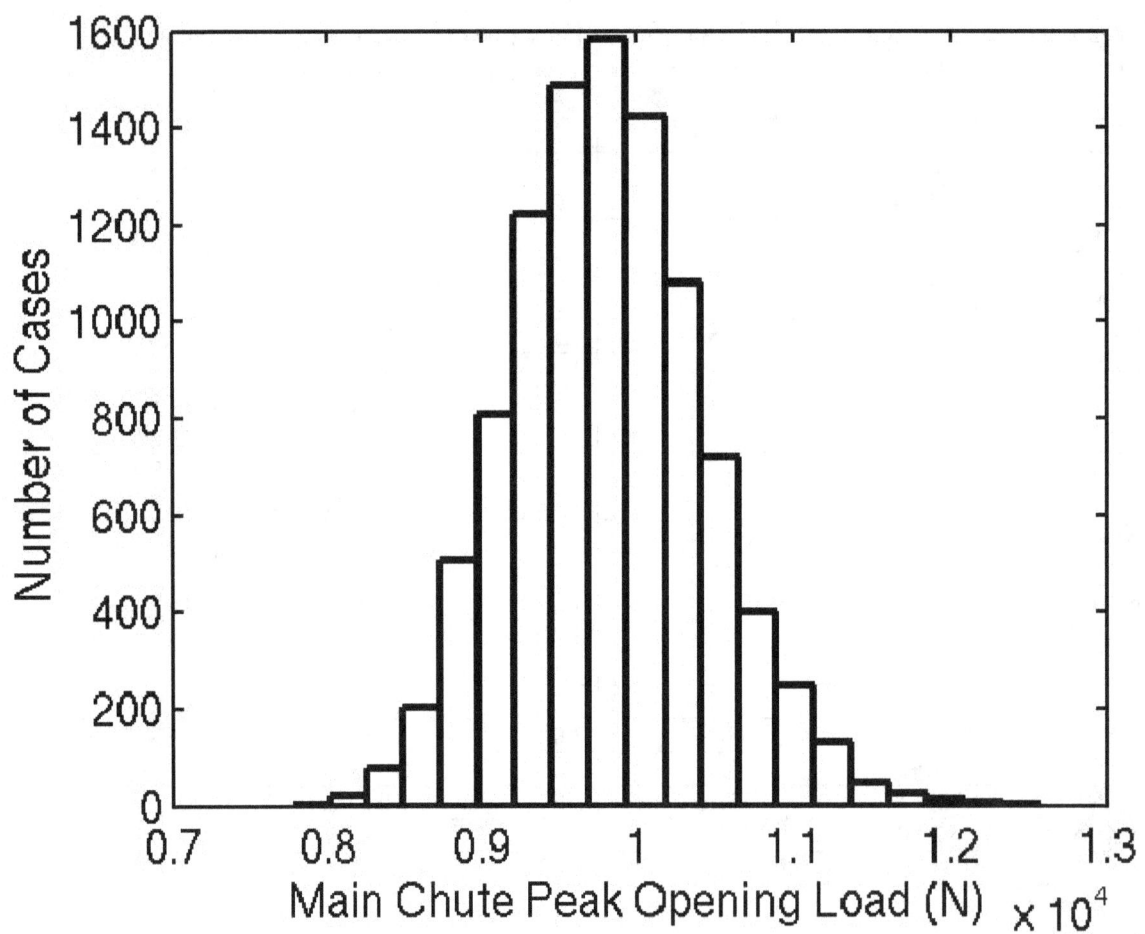

Figure F-1. Main Parachute Peak Opening Load using ESA Parachute Drag Model

NASA Engineering and Safety Center Technical Assessment Report	Document #: RP-05-67	Version: 1.0
Title: **Independent Technical Assessment of Cassini/Huygens Probe Entry, Descent and Landing (EDL) at Titan**		Page #: 77 of 116

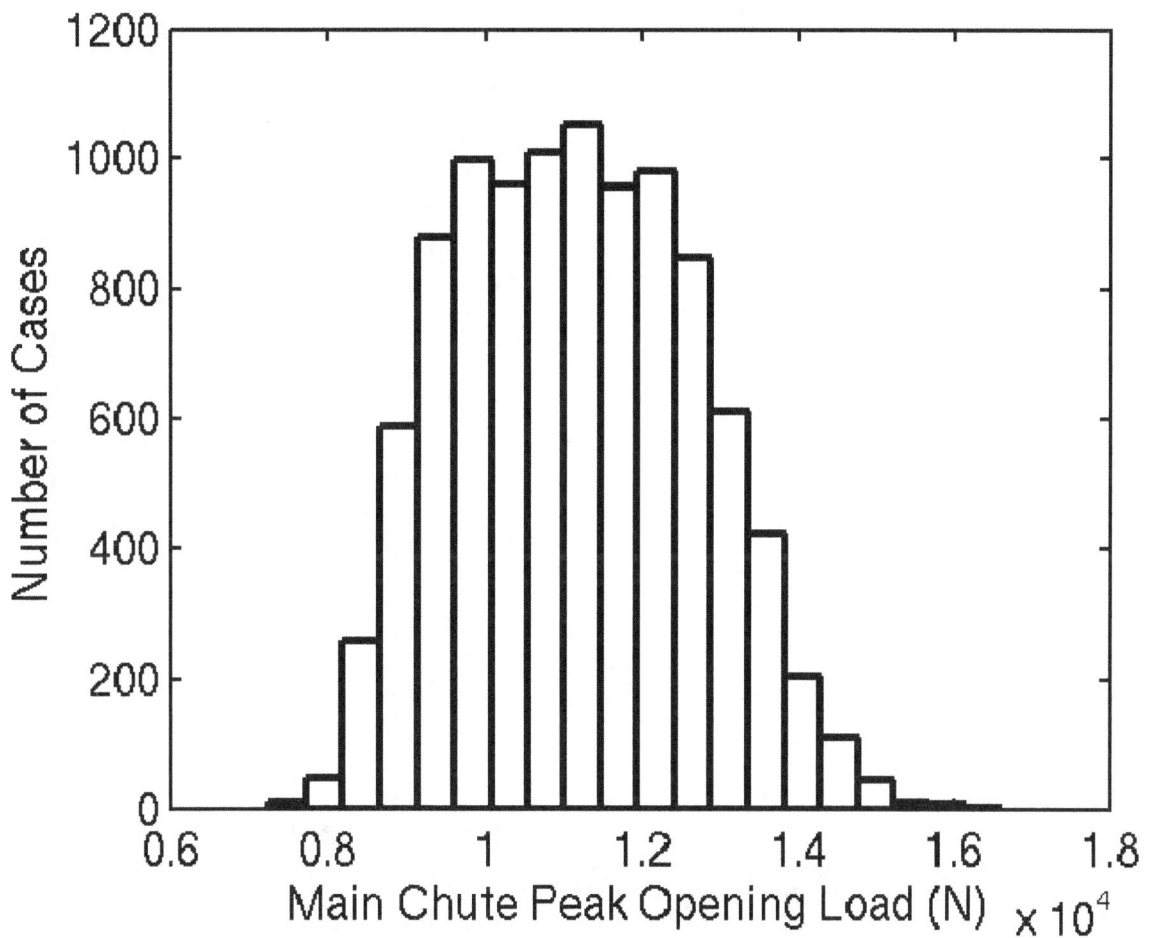

Figure F-2. Main Parachute Peak Opening Load using NASA 1 Parachute Drag Model

F.3.2 Total Time-Of-Flight

The design requirement time-of-flight (less than 2.5 hours from pilot parachute deployment to impact) is exceeded in less than 1 percent of the Monte Carlo runs using the Vorticity 1 parachute drag model set. When using the NASA 2 parachute drag model set, the design requirement time-of-flight is exceeded by approximately 25 percent of the Monte Carlo runs. (Note that these runs used the TA average atmosphere profile with Titan-GRAM, as discussed in Appendix D.) The differences in these results have been traced to the parachute drag models as discussed in Appendix C. Figures F-3 and F-4 show the time from pilot parachute deployment to the impact when using the Vorticity 1 and NASA 2 parachute drag model sets, respectively.

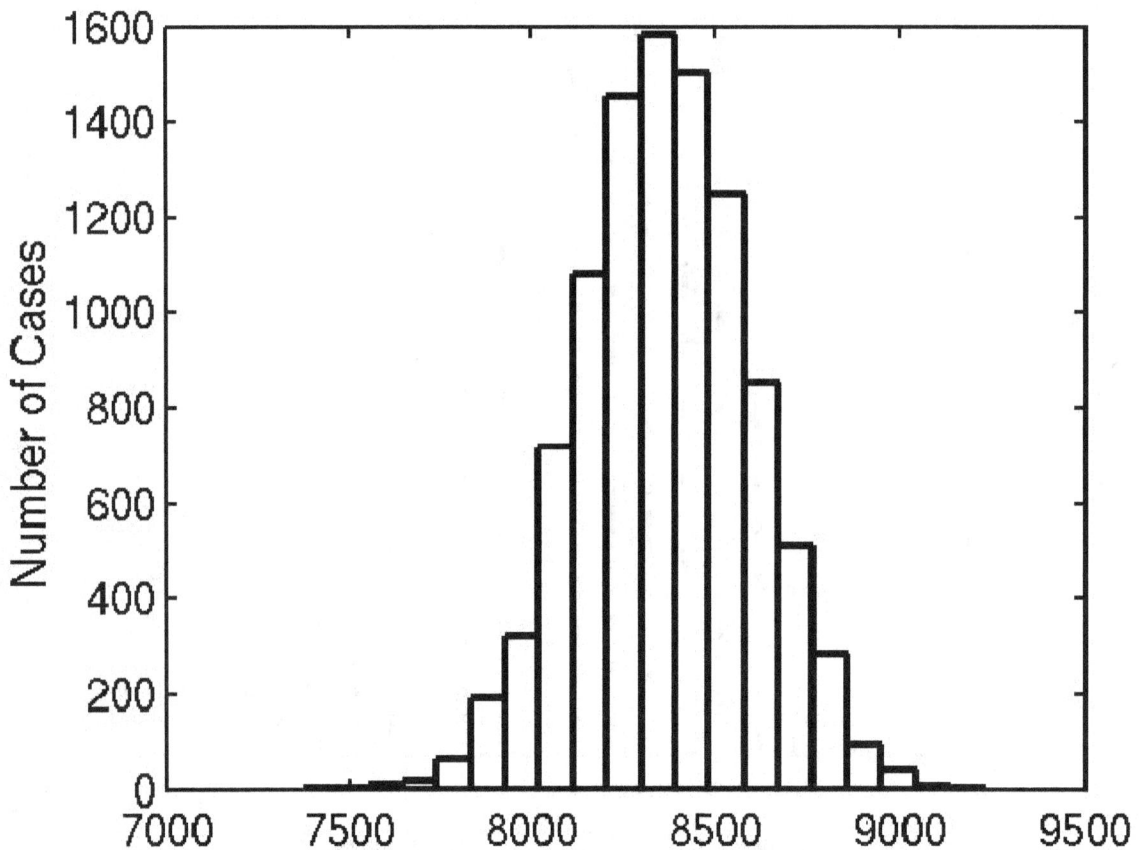

Figure F-3. Time of Impact from Pilot Parachute Deployment using Vorticity 1 Parachute Drag Model

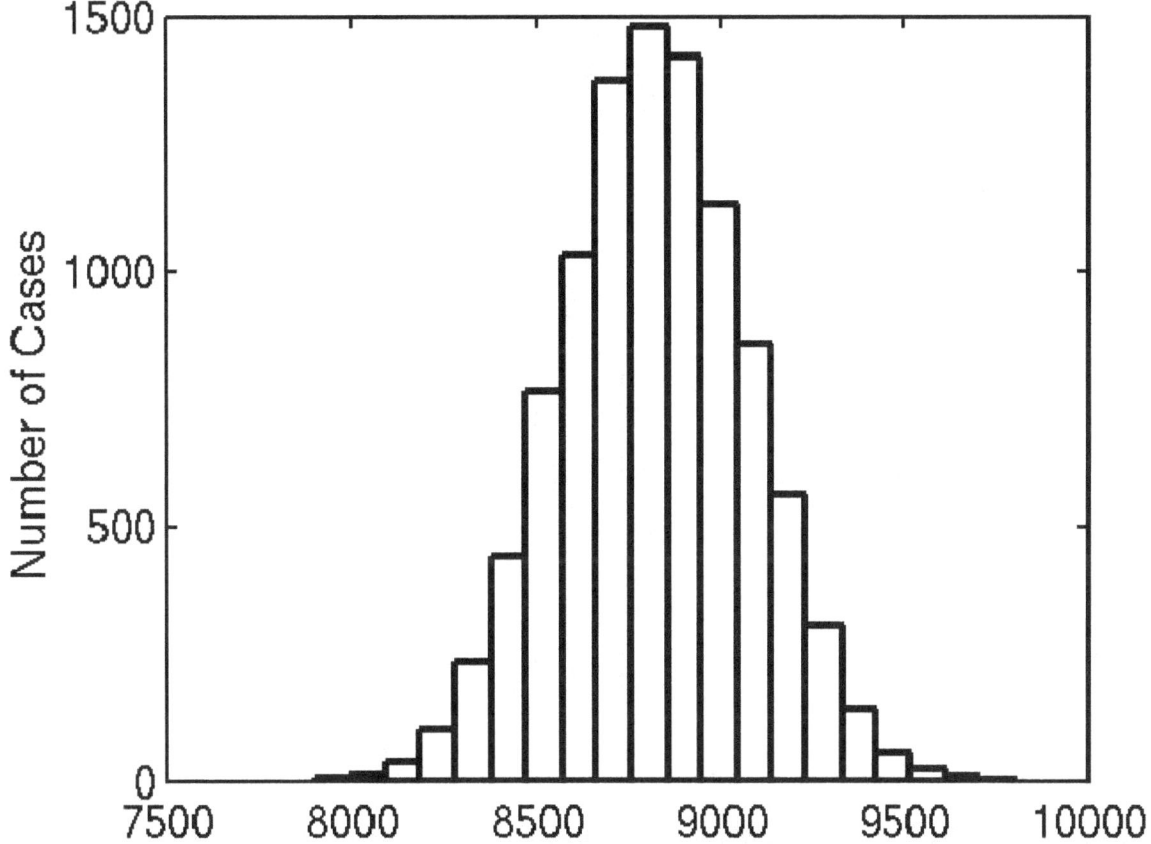

Figure F-4. Time of Impact from Pilot Parachute Deployment using NASA 2 Parachute Drag Model

F.3.3 Aerothermal Indicator Bounding Cases

The radiative and turbulent convective components of the aerothermodynamic environment appear to have been underestimated by the ESA during the Huygens probe design, and the uncertainties applied in ESA's analysis were considerably less than those that normally would be applied by NASA during the design process. Several stressing cases were identified from Monte Carlo analyses using aerothermal indicators. Figure F-5 shows the results from a Monte Carlo run. From these results, the 99.87% high case was identified. Trajectories for these cases were then supplied to the aerothermodynamic team for more detailed CFD analyses. This procedure was applied and repeated for several Monte Carlo cases.

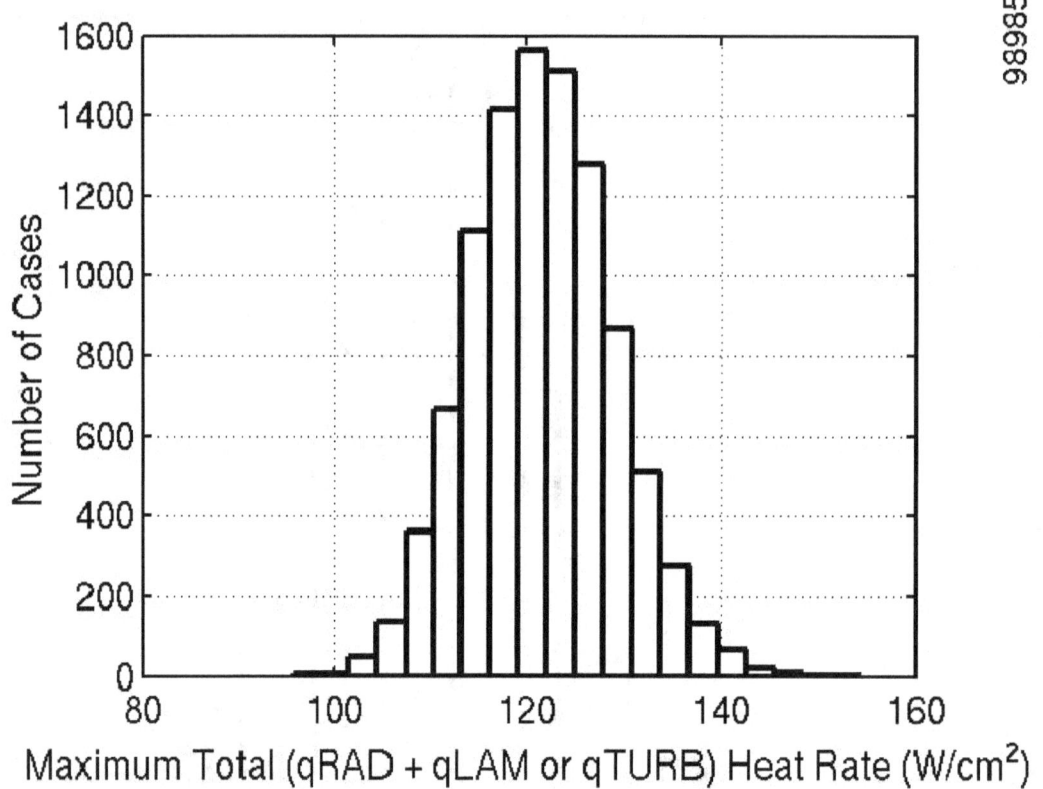

Figure F-5. Maximum Heat Rate from Monte Carlo Output

F.3.4 High Angle-of-Attack at Peak Heating Case

The ESA analysis assumed that the Huygens probe angle-of-attack at peak heating would be zero. The NASA Monte Carlo results showed that angles-of-attack as high as five degrees (3-sigma) are possible. Figure F-6 shows maximum total angle-of-attack at maximum heat rate for a Monte Carlo run using the latest versions of all the models. Using the trajectory from this 3-sigma case, the aerothermodynamic environment analysis indicated that this high angle-of-attack at peak heating would produce some heating amplification when compared to the ESA assumed attitude at peak heating. However, the NASA team concluded that the amount of heating amplification would not raise the risk above the low-to-moderate range.

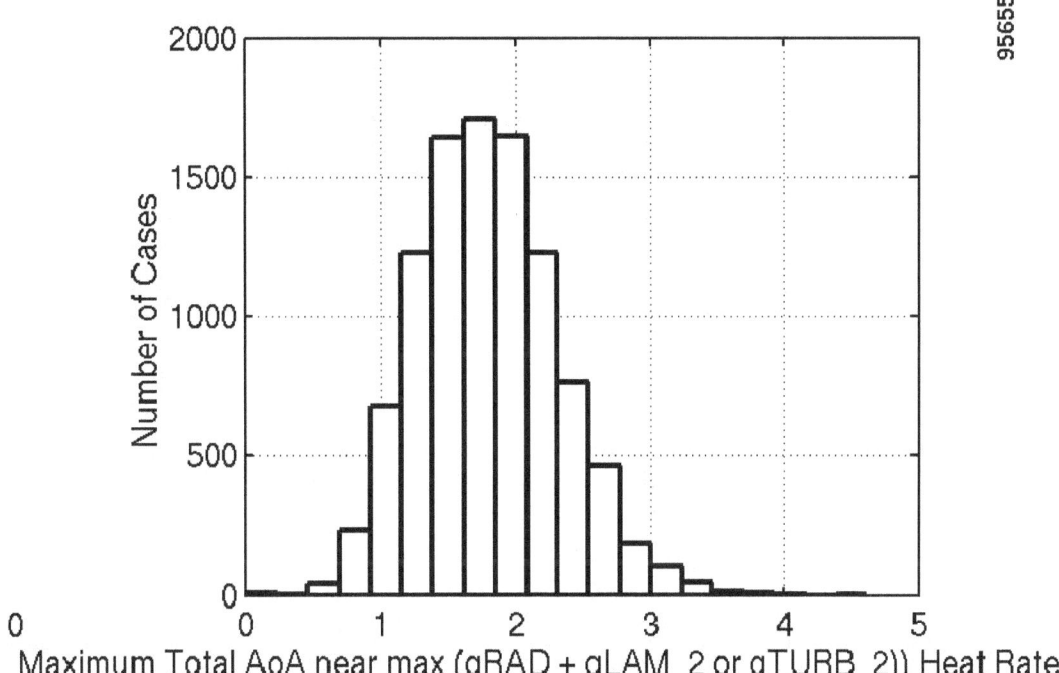

Figure F-6. Maximum Total Angle-of-Attack at Maximum Heat Rate

F.3.5 Entry Sensitivity Assessment

Various mission parameter sensitivity assessments were completed for the Huygens probe entry to determine the potential for modifying maximum heating rate, heat load, angle-of-attack at atmospheric interface, time of flight, etc. These assessments included changing the mean flight path angle at atmospheric interface, reducing the flight path angle uncertainty at atmospheric interface, and varying the nominal probe target orientation at atmospheric interface. All these assessments showed only small sensitivities. These insensitivities were traced to the large atmospheric density scale height of Titan (~40 km as compared to ~7km for Mars and Earth).

NASA Engineering and Safety Center Technical Assessment Report	Document #: RP-05-67	Version: 1.0
Title: Independent Technical Assessment of Cassini/Huygens Probe Entry, Descent and Landing (EDL) at Titan		Page #: 82 of 116

Appendix G. Aerothermodynamics

Prepared by
Brian Hollis and Michael Wright
NASA LaRC and NASA Ames

G.0   Introduction

G.1   Background

The aerothermodynamic environment is defined as the convective (both laminar and turbulent) and radiative heat-transfer rates and time-integrated heat-transfer loads to which a vehicle is exposed. This environment determines both the selection of and the thickness of the Thermal Protection System (TPS) material required to prevent the interior of the spacecraft from overheating. In the analysis of the Huygens entry, the two main aerothermodynamic technical challenges were in the modeling of the radiative heating and turbulent convective heating components.

1. Prediction of the radiative heating was a technical challenge that is beyond what is considered the validated, state-of-the-art methodology currently in use by NASA. For this mission, the radiative heating component of the total heating was of equal magnitude to the convective heating. This result differed from the experience of Earth and Mars entries at similar velocity, for which radiation would be negligible. The radiation at Titan is primarily due (estimated to be 80% to 95% for Huygens) to the specie CN that is produced from Titan's N_2-CH_4 atmosphere by high-temperature reactions in the bow-shock wave of the vehicle. While NASA has conducted missions to both Venus and Jupiter in which large radiative rates were produced, the uncertainty levels were very high due to a lack of suitable ground or flight based validation data for strongly radiating flows.

2. The prediction of boundary-layer transition onset is a difficult problem that is both configuration (vehicle shape and TPS selection/layout) and destination (planetary atmosphere) dependent. Without time for a thorough investigation of transition behavior for Huygens, a conservative transition criteria (boundary-layer parameter Re_θ/M_e exceeding 200 along the vehicle surface) was applied which predicted turbulent boundary-layer flow, with accompanying turbulent heating augmentation, over much of the trajectory.

The aerothermodynamic methodology employed in the design of the Huygens heat shield was raised as an issue during the ESA delta-FAR review in January 2004. Areas of concern were identified by the NESC team and discussed at the November 4-5, 2004 TIM, as follows:

- The magnitude of the ESA-predicted radiative heat flux appeared to be low in comparison to NASA models. The NESC team felt that the methodology employed by ESA to predict the CN radiation during the design of Huygens was incomplete and potentially non-conservative in that a finite-rate excitation model (as opposed to an equilibrium Boltzmann distribution) for the CN molecule was employed that lacked some of the important radiative processes and had not been validated (e.g. through comparison to shock tube data). This was an issue both because the actual heat load could exceed the design limit of the TPS material and because of the possibility that a large amount of short wavelength radiation (CN radiates primarily in the violet portion of the spectrum) could potentially penetrate the low-density TPS material and cause subsurface heating and spallation.

- The species diffusion model employed in the ESA CFD simulations resulted in non-conservative predictions of both laminar and turbulent convective heat transfer. The ESA model did not incorporate a multi-component diffusion model, as did the NESC CFD models, but instead employed a simpler binary-diffusion model which was suitable for modeling the behavior of Earth's atmosphere, but not Titan's.

- ESA concluded that boundary-layer transition would not occur until well after peak heating on the trajectory and, thus, that turbulent heating augmentation would not be a significant factor. ESA baseline estimates for transition onset were based in part on ARD flight data (Johnston 2002), but the utility of these data as well as their applicability to Huygens was questionable. Furthermore, recent ground tests (Hollis et al 2005, Olejniczak et al 2005, Wright et al 2005) performed by members of the NESC team in support of the MSL and Titan aerocapture missions indicated that transition could occur much earlier than estimated by ESA.

G.2   Methodology

The aerothermodynamic analysis of Huygens entry into Titan's atmosphere was conducted by separate teams from NASA's Langley and Ames Research Centers, operating in a collaborative framework, but employing completely independent tools sets to best identify modeling inconsistencies and bound the large uncertainties presented by this challenging problem.

Computational Fluid Dynamics simulations of the Huygens flow field were performed using the LAURA code (Gnoffo, 1990) at Langley and using the DPLR code (Wright, 1998) at ARC. Both codes have been extensively employed over the last decade in both the development of

NASA Engineering and Safety Center
Technical Assessment Report

Document #:
RP-05-67

Version:
1.0

Title:
Independent Technical Assessment of Cassini/Huygens Probe Entry, Descent and Landing (EDL) at Titan

Page #:
84 of 116

actual planetary missions (e.g. Genesis, Stardust, MER, Mars Science Laboratory) and in systems analysis studies which were conducted to identify challenges for future missions (e.g. a Titian aerocapture study conducted in 2002-2003). Both codes employ shock-capturing, finite-volume, thermo-chemical, non-equilibrium formulations to simulate hypersonic flow fields.

Radiative heating computations were performed by Langley using the RADEQUIL code (Nicolet, 1970) and by ARC using the NEQAIR code (Whiting 1996). The RADEQUIL code employs rapid, approximate techniques for radiation modeling, while the NEQAIR code employs a detailed, but more computationally intensive method. In both methods, the net radiative heat transfer to the vehicle surface is computed by a one-dimensional integration of the radiative emission and absorption from the shock layer to the surface. This 1D result is then multiplied by a factor to account for surface curvature. The value of this factor was set at 0.75 based on detailed analysis of the radiative view factors for selected cases. As detailed in the results of the Titan aerocapture systems study (Olejniczak 2003, Takashima, 2003), in which the members of the NESC team participated, one of the greatest challenges in the prediction of radiation at Titan is the coupling between the radiation transport computation and the flow field computation. This is required for a physically correct solution to this problem because the large amount of energy converted to radiation in the shock layer must be accounted for in the overall energy balance employed in the flow field computation. At present, a detailed, tightly-coupled procedure for an absorbing shock layer (i.e., one in which some of the emitted radiation is captured by the gas between the shock and the vehicle) is beyond current NASA capabilities. A fully-coupled procedure for the special case of an optically-thin (non-absorbing) gas has been implemented between the DPLR and NEQAIR codes (Wright et al. 2005 JTHT). However, because significant absorption was predicted for the Huygens entry conditions, the default procedure for coupling radiative energy loss to flow field modeling was an approximate engineering correction (Goulard, 1961; Tauber, 1971) to the uncoupled methodology. This method was originally developed for air or gas-giant (H_2-He) radiation, rather than CN radiation. The optically-thin coupling capability of the DPLR-NEQAIR codes was employed to better tailor the parameters used in the Tauber-Wakefield approximation for Titan entries.

Physical models in the codes are based on a combination of NASA best-practices employed for other entry simulations and Titan-specific enhancements. The chemical kinetics for Titan's atmosphere were taken from Gokcen et al (2004). The flow field was assumed to be in thermo-chemical non-equilibrium with separate translation and vibrational-electronic temperatures as per the two-temperature models of Park (1989) and Gnoffo (1989). Multi-component species diffusion was modeled using the methods detailed by Sutton (1998) in LAURA and Ramshaw (1990) in DPLR. These two approaches are nearly identical in practice, and both have been shown to accurately model true multi-species diffusion. The surface temperature was assumed to be in radiative equilibrium and the surface to be fully-catalytic. Turbulent simulations were computed using the compressible Baldwin-Lomax (1978) turbulence model with compressibility corrections (Cheatwood and Thompson 1993). For radiation modeling, an equilibrium

NASA Engineering and Safety Center Technical Assessment Report	Document #: RP-05-67	Version: 1.0
Title: **Independent Technical Assessment of Cassini/Huygens Probe Entry, Descent and Landing (EDL) at Titan**		Page #: 85 of 116

(Boltzmann) assumption was employed, in which the excited states of the CN molecule were assumed to maintain an equilibrium distribution at the mixture vibrational-electronic temperature as computed in the CFD.

Aerothermodynamic computations were performed along several trajectories, beginning with ESA estimates for the maximum heat-rate and heat-load trajectories, then progressing through two revisions of max-rate and max-load trajectories generated by the NESC team using the POST code with independently-generated aerodynamic databases and the most up-to-date Titan atmospheric models. The final results and recommendations presented to the NESC review panel were based on the trajectory identified as worst case (99.7%) heat-rate and heat-load trajectories generated on November 11, 2004 (shown in Figure G-1).

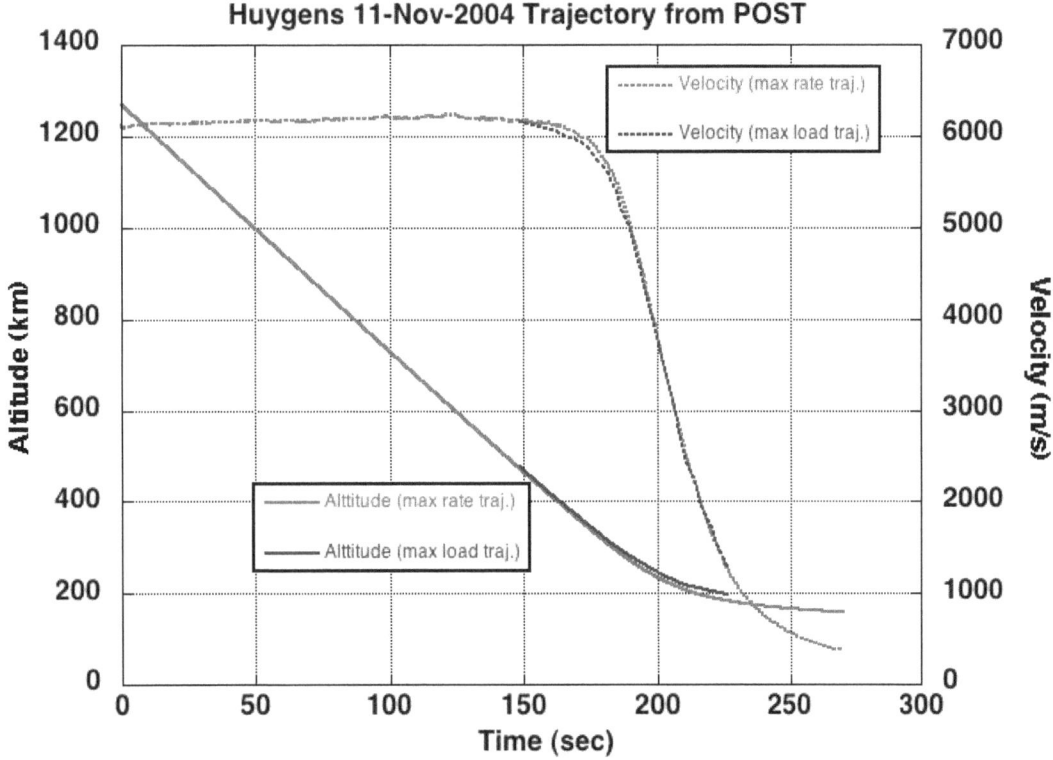

Figure G-1. 11-Nov-2004 Max Heat-Rate and Max Heat-Load Trajectories from POST

Multiple solutions were computed on each trajectory (between 5 and 9) and then curve-fits and/or table-lookup functions were generated from the predicted heating rates as a function of time along the trajectory for various locations on the heat shield. Sample comparisons for the convective, radiative and total heat-transfer rates from the ARC (DPLR/NEQAIR) and LaRC (LAURA/RADEQUIL) predictions for at the nose of the vehicle are shown in Figures G-2 through G-4. The time-history curve-fits and lookup tables were then fed back into the POST trajectory simulation to obtain more accurate Monte Carlo determination of the worst-case heat-rate and heat-load trajectories.

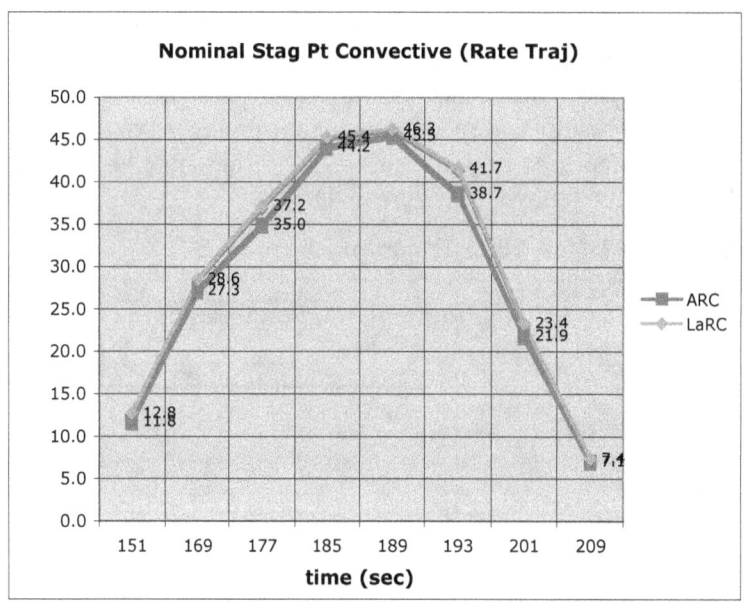

Figure G-2. Nominal Convective Heating Time-History at Stagnation Point on Rate Trajectory

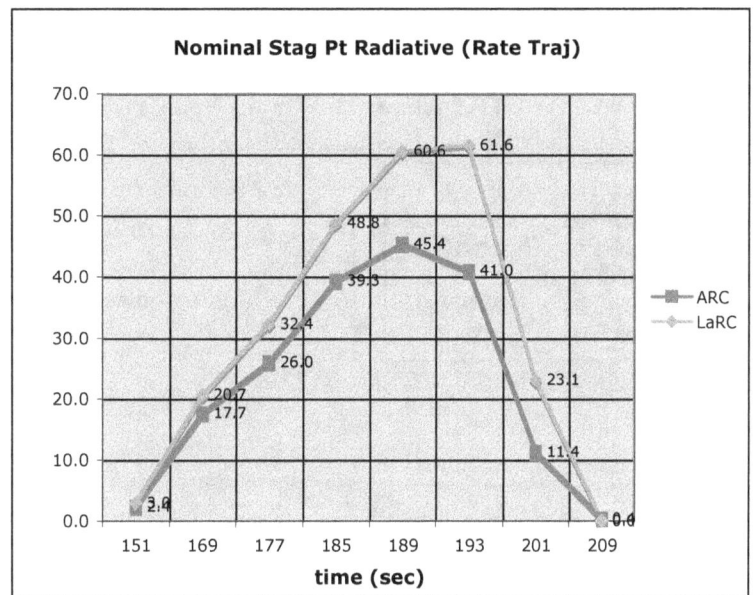

Figure G-3. Nominal Radiative Heating Time-History at Stagnation Point on Rate Trajectory

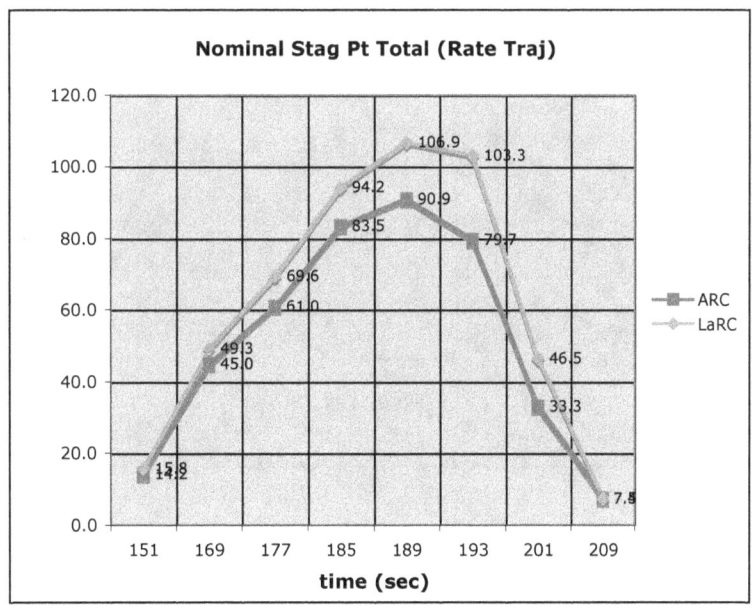

Figure G-4. Nominal Total Heating Time-History at Stagnation Point on Rate Trajectory

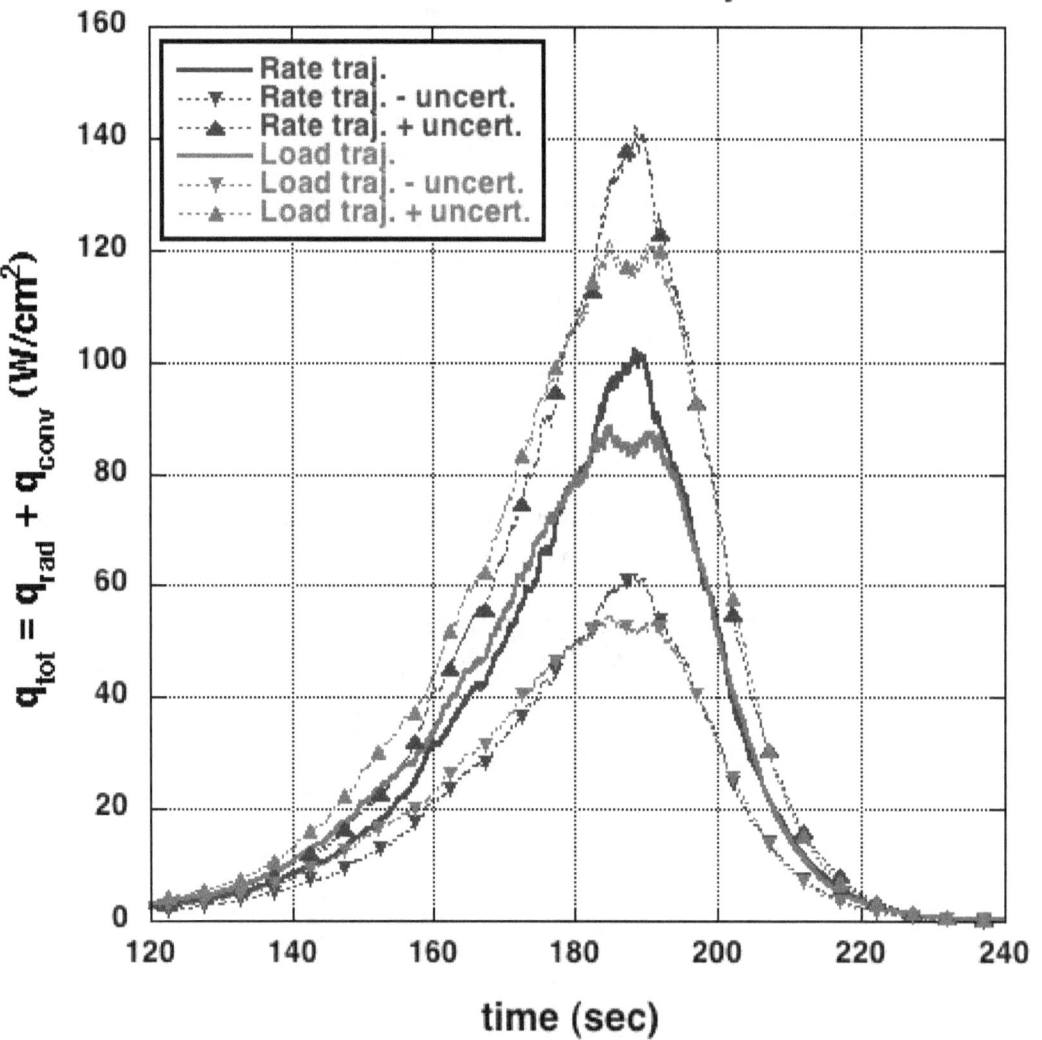

Figure G-5. Heat-Rate at Stagnation Point

Time histories of the heat-rate and heat-load along the final November 14th trajectory at the nose (stagnation point) and at the conical-section midpoint are presented in Figures G-5 through G-8. The baseline heating rates and loads in these figures represent an average of the LaRC (LAURA/RADEQUIL) and ARC (DPLR/NEQAIR) predictions along the trajectory. DPLR and LAURA predictions for convective heating generally agreed to within 10%, but larger discrepancies existed between the predictions of radiative heating. Due to the short turnaround time of this analysis some of the reasons for this discrepancy are still not known, but most of the difference appears to be due to the predicted amount of radiation absorption in the shock layer.

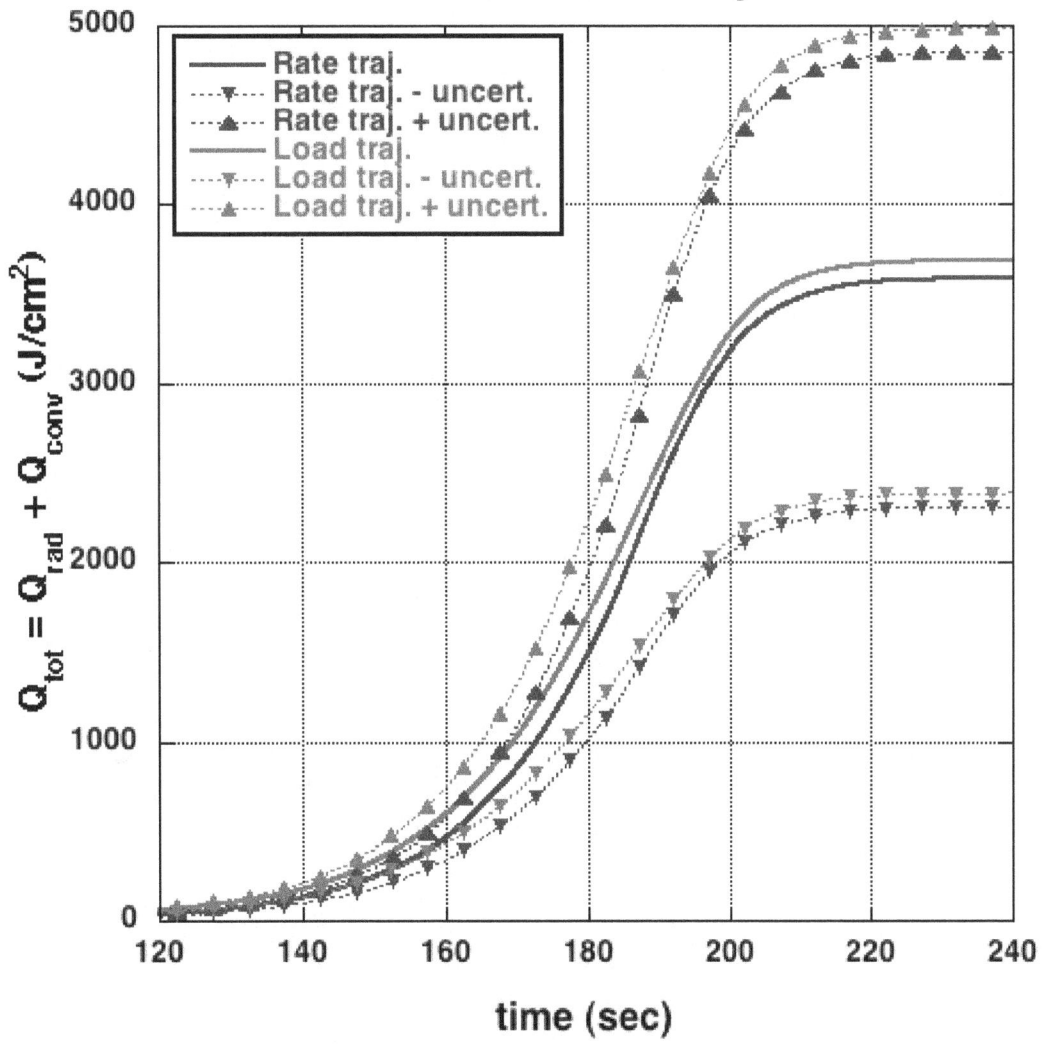

Figure G-6. Heat-Load at Stagnation Point

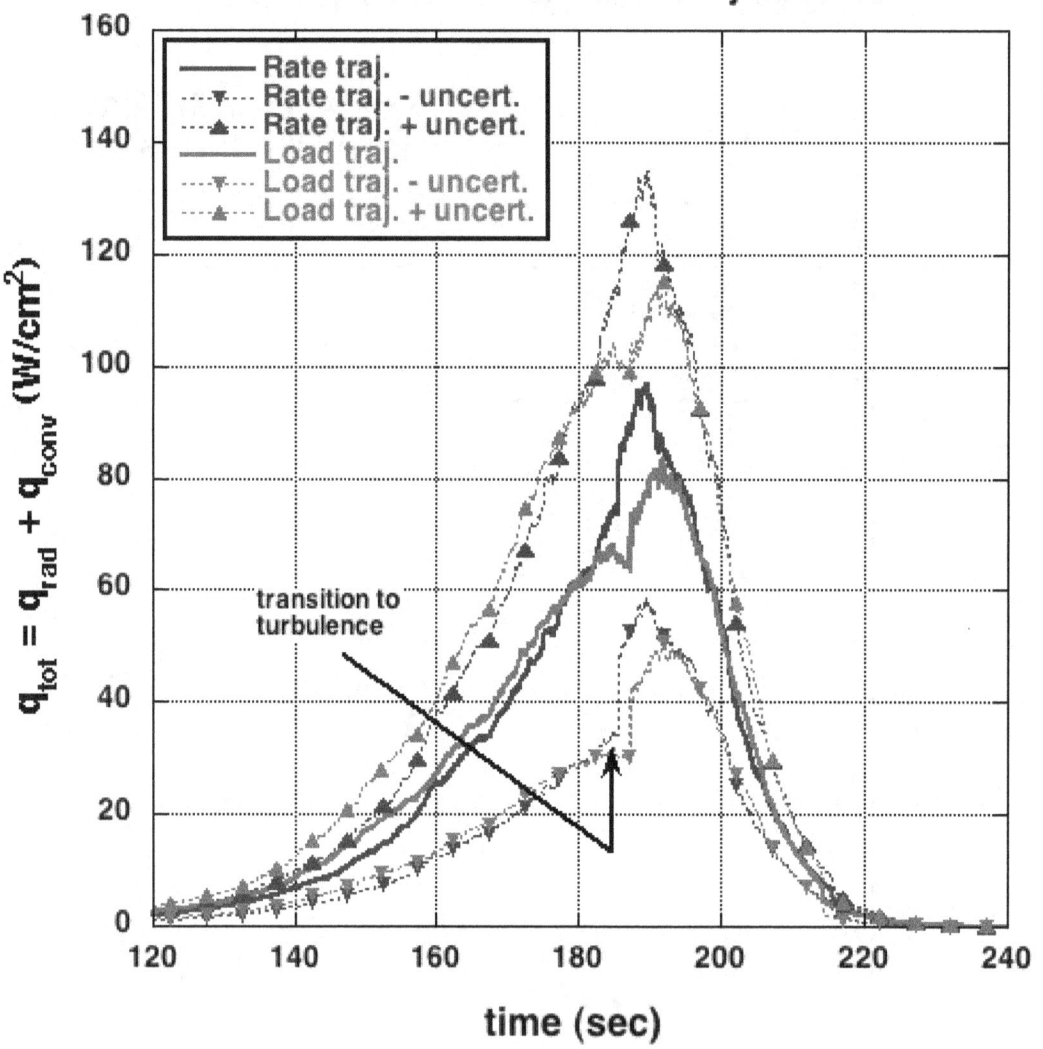

Figure G-7. Heat-Rate at Mid-Cone

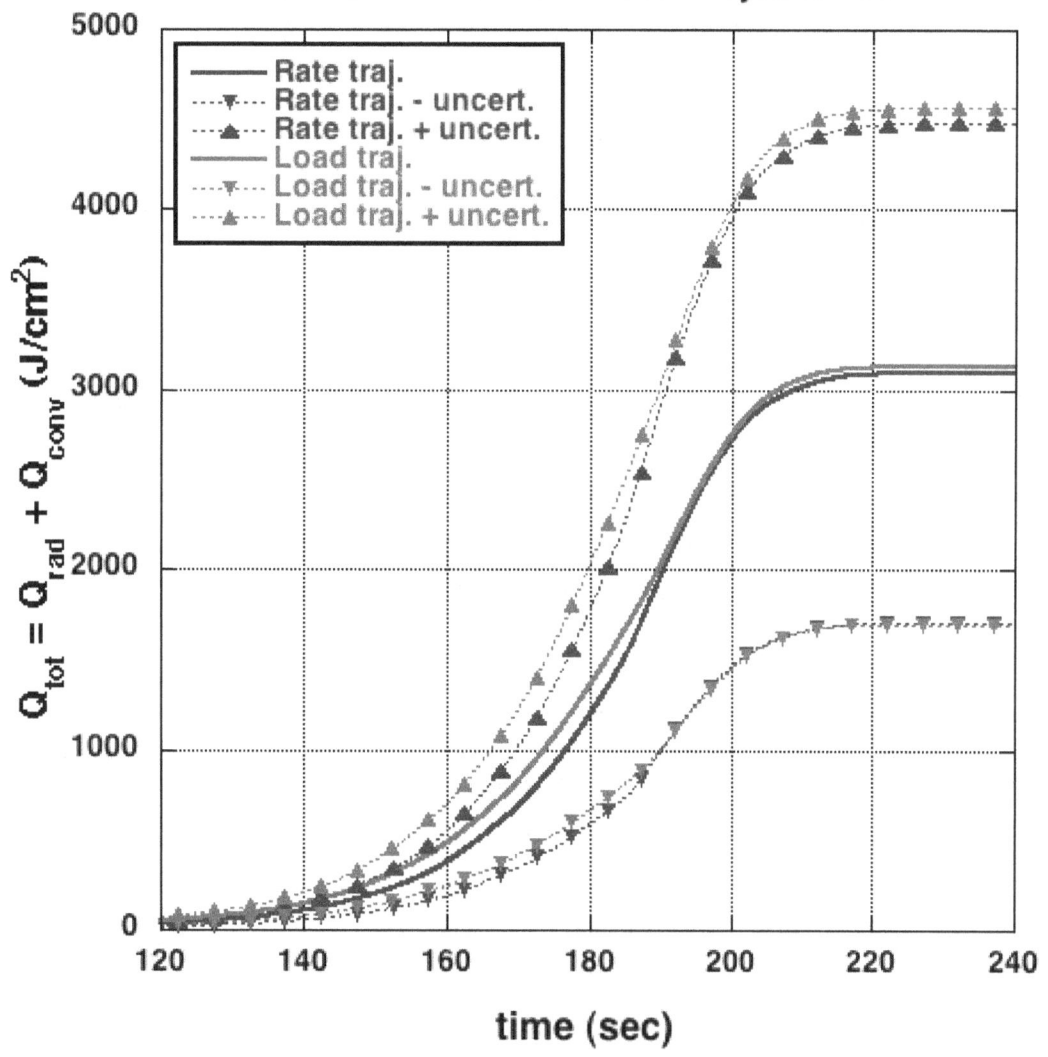

Figure G-8. Heat-Load at Mid-Cone

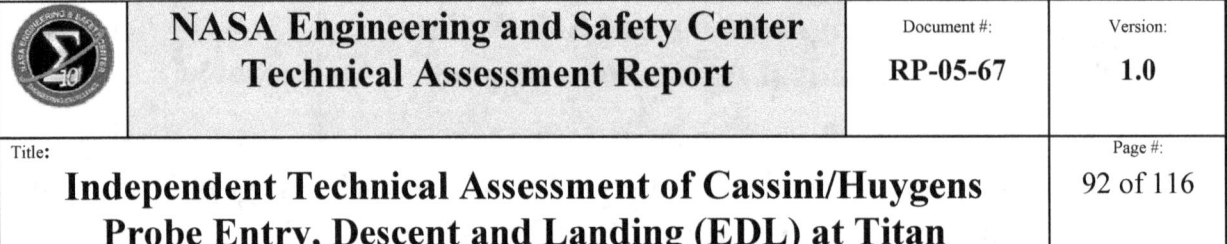

NASA Engineering and Safety Center Technical Assessment Report	Document #: RP-05-67	Version: 1.0
Title: **Independent Technical Assessment of Cassini/Huygens Probe Entry, Descent and Landing (EDL) at Titan**	Page #: 92 of 116	

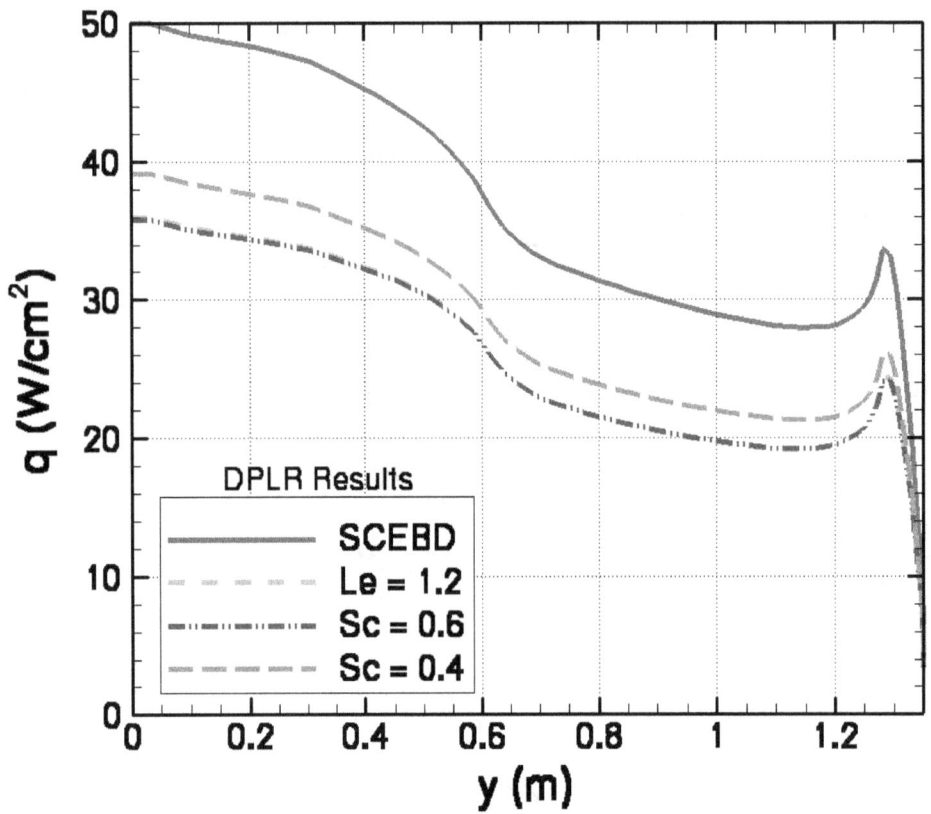

Figure G-9. Example of Diffusion-Modeling Effects on Convective Heating

Uncertainty bounds are also presented in these figures. Due to the limited time in which the Huygens review was conducted, the determination of these uncertainty bounds was much less rigorous than current best practices (e.g. the MSL design process) would require, and should be taken as conservative "best-guesses". The breakdown of uncertainties is as follows: ±15% for laminar, convective heating predictions; ±40% for turbulent heating predictions; and ±60% for radiative heating predictions. The high uncertainty in the radiation predictions is based on the limited knowledge of the radiative properties of the CN molecules, the limited knowledge of the concentration of CH_4 in Titan atmosphere (which has a first order effect on the amount of CN formed and thus on the radiative heating levels) and the fact that only an approximate radiation flow field coupling methodology could be employed in the analysis. The important point to note in this set of figures is that when the positive uncertainties were added to the integrated heat-load, the resulting values exceeded 4000 J/cm^2 at both the stagnation point and mid-cone point.

	NASA Engineering and Safety Center Technical Assessment Report	Document #: RP-05-67	Version: 1.0
Title: **Independent Technical Assessment of Cassini/Huygens Probe Entry, Descent and Landing (EDL) at Titan**			Page #: 93 of 116

This level was stated to be beyond the forebody TPS design limits by ESA; in other words, the NESC analysis indicated the possibility of heat shield failure.

G.3   Key Results

Accurate modeling of species diffusion was found to play a large role in both laminar and turbulent convective heating. Diffusion modeling is particularly important for Titan due to the CH_4 in the free stream, which quickly dissociates into its constituents after passing through the shock wave. This dissociation leads to a large number of light hydrogen atoms in the post-shock gas, which diffuse rapidly as compared to the heavier N_2 atoms and molecules. Under these conditions, a constant Schmidt number diffusion model (such as employed by ESA) is inaccurate. Both LAURA and DPLR contain accurate diffusion models, which predict up to a 40% increase in convective heating rate over the constant Schmidt number approach (an example is shown in Figure G-9).

Although most of the analysis was performed for the nominal $\propto = 0°$ entry, a limited number of cases were also run at small angles-of-attack ($\propto = 2\text{--}5°$) in order to assess the effects of predicted oscillations near peak heating. This small angle-of-attack had minimal impact on the convective (laminar and turbulent) heating rate. The effect on the radiative heating rate was more significant, with the wind-side heating increasing and the leeside decreasing with $\propto$. However, given that in flight the vehicle would be oscillating about $\propto = 0°$, the net effect of these oscillations on the integrated heat load was determined to be small.

The baseline ESA analysis for the aftbody (i.e., the payload) assumed that heating rates were a percentage of the forebody stagnation convective and radiative values. Values of up to 14% of the stagnation point radiative heating were used at certain places on the aftbody. Fairly late in the analysis (a week prior to the November 4-5, 2004 Paris meeting) it was learned that the ESA-estimated margins on the aftbody TPS were actually smaller than those on the forebody for many of the TPS sizing trajectories. Given the short time to respond to this information, a full analysis of the aftbody heating rates was not possible. However, a limited number of calculations were performed on an ESA-provided design trajectory. This work was based on previous analysis of Titan base heating for the ISP aerocapture mission (Olejniczak 2004). Based on this limited analysis it was determined that the ESA estimates of aftbody convective heating were reasonable, but that the estimates of aftbody radiative heating were likely conservative (over predicted the true radiative heating rates) by a factor of 4 or more. This result is qualitatively consistent with flight data from Fire-II and Apollo 4 and 6, which carried radiometers on the aftbody and failed to measure any significant signal.

The dominant contributor to the radiative heating was the CN molecule, but other radiators, including N_2, N, C_2, and C also contributed to the net heat flux, especially early in the entry. All of these species were considered in the final analysis. The non-adiabatic effect of flow field radiation coupling was found to be significant throughout the entire entry, reducing the peak

NASA Engineering and Safety Center Technical Assessment Report	Document #: RP-05-67	Version: 1.0
Title: **Independent Technical Assessment of Cassini/Huygens Probe Entry, Descent and Landing (EDL) at Titan**		Page #: 94 of 116

radiative heat flux by nearly a factor of two. As discussed earlier, the baseline radiative heating analysis was conducted assuming a Boltzmann distribution of excited states, primarily because a validated radiation model for CN was not available. However, it was anticipated that such an assumption would provide a conservative estimate of the total radiative heat transfer rate. Concurrently with the Huygens analysis, a new radiation model for CN was under development at NASA ARC, using shock tube radiation data obtained in 2004 under the ISP Program. Unfortunately, this analysis was not completed in time to directly influence the models used in this work, but preliminary results obtained in early December 2004 and published in January of 2005 (Bose et al, 2005) indicated that the Boltzmann model was indeed conservative. Given the preliminary nature of this conclusion it was decided not to alter the risk assessment for the radiative heating environment, but the results did provide some confidence that there was conservatism in the design. The new nonequilibrium CN radiation model based on these data indicated that the Boltzmann assumption likely over-predicted the radiative heating for Huygens by at least a factor of 2.

G.4   Conclusions

NASA concerns in regard to the aerothermodynamic environment were presented during face-to-face meetings between the NASA aerothermodynamics team (comprised of representatives from LaRC and ARC) and ESA's aerothermodynamics team on November 4-5 2004. Aerothermodynamic models and assumptions (as detailed above) which represented the current best practices of the NASA team were proposed to, and accepted by, the ESA team as the standard by which the final, pre-release aerothermodynamic analysis would be conducted. The ESA agreed to implement, if possible in the limited time available, these practices in their numerical tools and use them to generate new heating rate and load estimates that would then be used as inputs to a thermal analysis of the integrity of the Huygens heat shield material. Ultimately, this implementation was not feasible before the release decision date. Instead, aerothermodynamic analyses performed by the NESC team using trajectories and atmospheric models, also generated by the NESC team, were employed as inputs to the ESA's thermal analysis.

The NESC team concluded that a conservative, worst-case analysis indicated the possibility that the integrated heat-load could exceed the ESA design specification of 4000 J/cm^2. However, it was concluded that this possibility did not present an unacceptable mission risk. This conclusion however, included two important caveats:

1.   The NESC aerothermodynamic analysis was conducted with less rigor than would be applied to a design phase NASA mission because the NESC review was authorized at such a late date (less than four months before the probe release decision). Potentially important issues which were not addressed in detail include the aftbody aerothermodynamic environment (only minimal computations were performed for the aftbody/wake flow); the possibility of heating augmentation and early transition onset

due to gaps or steps between the tiles of the Huygens forebody heat shield; the effects of deviations in angle-of-attack from the nominal 0° on heating; the lack of high-fidelity coupling between flow field and radiative transport computations (approximate engineering correlation were substituted), and ground-test validation of the employed model for CN radiative heating (data fortuitously taken for a different program could not be reduced in time to be employed).

2.   The definition of acceptable risk as applied to a mission that was already in flight and for which very little flexibility in the trajectory profile was available (as detailed separately) was less stringent than would be applied to a new NASA mission.

References

Johnston, I. A., Weiland, M., Schramm, J. M., Hannemann, K., Longo, J., "Aerothermodynamics of the ARD: Postflight Numerics and Shock-Tunnel Experiments," AIAA 2002-0407, January 2002.

Hollis, B. R., Liechty, D. S., Wright, M. J., Holden, M. S., Wadhams, T. P., MacLean, M., and Dyaknov, A., "Transition Onset and Turbulent Heating Measurements for the Mars Science Laboratory Entry Vehicle," AIAA Paper 2005-1437, January 2005.

Olejniczak, J., Wright, M. J., Laurence, S., and Hornung, H. G., "Computational Modeling of T5 Laminar and Turbulent Heating Data on Blunt Cones, Part 1: Titan Applications," AIAA Paper 2005-0176, January 2005.

Wright, M. J., Olejniczak, J., Brown, J. L., Hornung, H. G., and Edquist, K. T., "Computational Modeling of T5 Laminar and Turbulent Heating Data on Blunt Cones, Part 2: Mars Applications", AIAA Paper 2005-0177, January 2005.

Gnoffo, P. A., "An Upwind-Biased, Point-Implicit Algorithm for Viscous, Compressible Perfect-Gas Flows," NASA TP-2953, February 1990.

Wright, M. J., Candler, G. V., and Bose, D., "Data-Parallel Line-Relaxation Method for the Navier-Stokes Equations," *AIAA Journal*, Vol. 36, No. 9, 1998, pp. 1603-1609.

Nicolet, W. E., "Advanced Methods for Calculating Radiation Transport in Ablation-Product Contaminated Boundary Layers," NASA-CR-1656, September 1970.

Whiting, E. E., Yen, L., Arnold, J. O. and Paterson, J. A., "NEQAIR96, Nonequilibrium and Equilibrium Radiative Transport and Spectra Program: User's Manual," NASA RP-1389, December 1996.

NASA Engineering and Safety Center
Technical Assessment Report

Document #:
RP-05-67

Version:
1.0

Title:

Independent Technical Assessment of Cassini/Huygens Probe Entry, Descent and Landing (EDL) at Titan

Page #:
96 of 116

Olejniczak, J., Wright, M., Prabhu, D., Takashima, N., Hollis, B., Zoby, E., and Sutton, K., "An Analysis of the Radiative Heating Environment for aerocapture at Titan," AIAA Paper No. 2003-4953, July 2003.

Takashima, N., Hollis, B., Zoby, E., Sutton, K., Olejniczak, J., Wright, M., and Prabhu, D., "Preliminary Aerothermodynamics Analysis of Titan Aerocapture Aeroshell," AIAA Paper No. 2003-4952, July 2003.

Wright, M. J., Bose, D., and Olejniczak, J., Impact of Flowfield-Radiation Coupling on Aeroheating for Titan Aerocapture, *Journal of Thermophysics and Heat Transfer*, Vol. 19, No. 1, January-March 2005, pp. 17-27.

Goulard, R., "The Coupling of Radiation and Convection in Detached Shock Layers," *Journal of Quantitative Spectroscopy and Radiative Heat Transfer*, Vol. 1, 1961, pp. 249–257.

Tauber, M. and Wakefield R., "Heating Environment and Protection During Jupiter Entry," *Journal of Spacecraft and Rockets*, Vol. 8, No. 3, 1971, pp. 630-636.

Gökçen, T., "N_2-CH_4-Ar Chemical Kinetic Model for Simulations of Atmospheric Entry to Titan," AIAA Paper 2004-2469, June-July 2004.

Sutton, K. and Gnoffo, P. A., "Multi-Component Diffusion with Application to Computational Aerothermodynamics," AIAA 1998-2575, June 1998.

Ramshaw, J. D., "Self-Consistent Effective Binary Diffusion in Multicomponent Gas Mixtures," *Journal of Non-Equilibrium Thermodynamics*, Vol. 15, No. 3, 1990, pp. 295-300.

Baldwin, B. S. and Lomax, H., "Thin Layer Approximation and Algebraic Model for Separated Turbulent Flow," AIAA Paper 78-257, January 1978.

Cheatwood, F. M., and Thompson, R. A., "The Addition of Algebraic Turbulence Modeling to Program LAURA," NASA TM-107758, April 1993.

Bose, D., Wright, M. J., Raiche, G., Bogdanoff, D., and Allen, G.A., "Modeling and Experimental Validation of CN Radiation Behind a Strong Shock Wave," AIAA Paper No. 2005-0768, January 2005.

Olejniczak, J., Prabhu, D. K., Bose, D., and Wright, M. J., "Aeroheating Analysis for the Afterbody of a Titan Probe," AIAA Paper 2004-0486, January 2004.

NASA Engineering and Safety Center Technical Assessment Report	Document #: RP-05-67	Version: 1.0
Title: Independent Technical Assessment of Cassini/Huygens Probe Entry, Descent and Landing (EDL) at Titan		Page #: 97 of 116

Gnoffo, P. A., Gupta, R. N., and Shinn, J. L., "Conservation Equations and Physical Models for Hypersonic Air Flows in Thermal and Chemical Nonequilibrium," NASA TP 2867, February 1989.

Park, C., "Assessment of Two-Temperature Kinetic Model for Air," *Journal of Thermophysics and Heat Transfer*, Vol. 3, No. 3, 1989, pp. 233-244.

NASA Engineering and Safety Center Technical Assessment Report	Document #: RP-05-67	Version: 1.0
Title: **Independent Technical Assessment of Cassini/Huygens Probe Entry, Descent and Landing (EDL) at Titan**		Page #: 98 of 116

Appendix H. Thermal Protection System

Prepared by
Bernard Laub and Michael Wright
NASA Ames Research Center

H.0   Introduction

H.1   Background

During FY 2002, a NASA team of specialists conducted a detailed systems analysis study of a potential aerocapture mission at Titan. During this study, it was learned that the primary heating during aerocapture (or entry) is radiation from the CN molecule formed in the shock layer. Furthermore, all of this radiation occurs at ultraviolet (UV) wavelengths, i.e., 350-420 nm. Bernard Laub, NASA Ames Research Center, who was leading the studies of thermal protection for this Titan aerocapture mission, expressed concern about how well proposed low-density TPS materials would perform when exposed to UV radiation. This concern was based on prior experience with laser irradiation of ablative composites that identified a trend of material performance degradation at shorter wavelengths due to in-depth radiant absorption. In the worst-case, this behavior leads to spallation of the material near the surface. Given that the most promising TPS candidates were low density, porous materials, this concern was identified as an uncertainty that could only be resolved through testing. Subsequently, since Mr. Laub was Principal Investigator on a project sponsored by the In-Space Propulsion (ISP) aerocapture project, it was decided to acquire and/or develop an appropriate test facility to evaluate the TPS candidates for exposure to UV radiation.

In November 2003, Jean-Pierre Lebreton, Huygens Mission Manager/Project Scientist, invited Mr. Laub to participate as a full member of the EDL panel for the Huygens Delta Flight Acceptance Review. Mr. Laub accepted and was given access to a broad range of Huygens documents describing the Huygens EDL design and more recent modifications incorporated after launch of the Cassini spacecraft. Panel members were encouraged to submit Review Item Discrepancy (RID) reports for consideration of the full panel. Mr. Laub raised the issue of whether the Huygens forebody TPS material, AQ60, had ever been tested with UV radiation and expressed his concerns.

During this review, ARC also requested the latest entry trajectory information from ESA and, upon receipt, did an independent analysis of the entry aerothermal environment for comparison with the environments used by the European Aeronautic Defence and Space Company (EADS) for TPS design. This comparison revealed significant differences between the ARC and EADS

predictions. This information was transmitted to the EDL Panel secretary, Thierry Blanquaert, who suggested a meeting prior to the Delta FAR Panel meetings scheduled for Cannes, February 2-4, 2004. This meeting was held at EADS Aquitaine on January 28, 2004.

At this meeting, hosted by Jean-Marc Bouilly (EADS), the issue of potential AQ60 semi-transparency and the differences between ARC and EADS entry heating predictions were discussed. It was concluded that both issues would be discussed during the full EDL panel meeting in Cannes.

At the Cannes meeting, NASA ARC offered to test AQ60 with the UV lamp facility being procured under the ISP activity. The panel recommended that this offer be accepted and the availability and schedule for delivery of test specimens was an issue left to be resolved between the ESA and EADS. In addition, the differences between ARC and EADS radiative heating were deemed so significant that the panel recommended direct discussion among experts at ARC, ESA and EADS.

NASA ARC defined desired TPS test specimen geometry, i.e., square specimens 75 x 75 mm x 20 mm thick, with each specimen having a 25 mm diameter plug for installation of in-depth thermocouples (to be done by NASA). EADS delivered eight specimens of AQ60 to NASA ARC in April 2004 for UV testing. Typical specimen geometry is shown in Figure H-1.

Figure H-1. AQ60 Sample Geometry for UV Radiation Tests

H.2   UV Lamp

To simulate the UV radiation heat flux for Titan aerocapture, a wide array of test facilities was investigated by NASA ARC during FY 2003, including solar energy concentrators, high-powered lasers, and several types of specialized lamps. The only device capable of providing the desired environment, i.e., simulating the predicted UV wavelength range, the total radiant heat load, and the peak radiant energy flux over a sufficient area for material testing was a mercury-xenon lamp — specifically a modification of a commercially-available mercury-xenon lamp. Mercury-xenon short-arc lamps are dosed with an exact amount of mercury in xenon gas at high pressure.

Figure H-2 shows the spectral distribution of the radiation predicted to the surface of the Titan probe in the near-UV and the spectral distribution for the test facilities considered for simulating the Titan probe's radiation heat flux. In Figure H-2, the intensity was scaled for clarity to separate out laser sources, broad spectrum lamps, the required shock layer spectra and the spectra available from the specific mercury-xenon lamp because the intensity is adjustable by tailoring the power of the source. Solar and xenon lamp radiation are dominated by the infrared wavelengths. In addition, the atmosphere filters out UV radiation so efficiently that earth-bound collectors could not provided sufficient UV power for testing. Several high-powered lasers were investigated, but the best wavelength matches were pulsed lasers, which are expected to produce misleading results for material screening. The mercury-xenon lamp spectrum overlaps well with the CN radiation from the Titan shock layer. The mercury-xenon lamp spectrum is dominated by the four main mercury lines and shows only traces from the broad, visible, and IR-dominated spectrum of xenon. The first three mercury lines are an excellent simulation of the predicted CN radiation at Titan. The fourth dominant line between 500 and 600 nm was easily filtered out, and a simple cold filter eliminated the visible and IR lines. Therefore, a unique mercury-xenon lamp was selected and acquired for these tests.

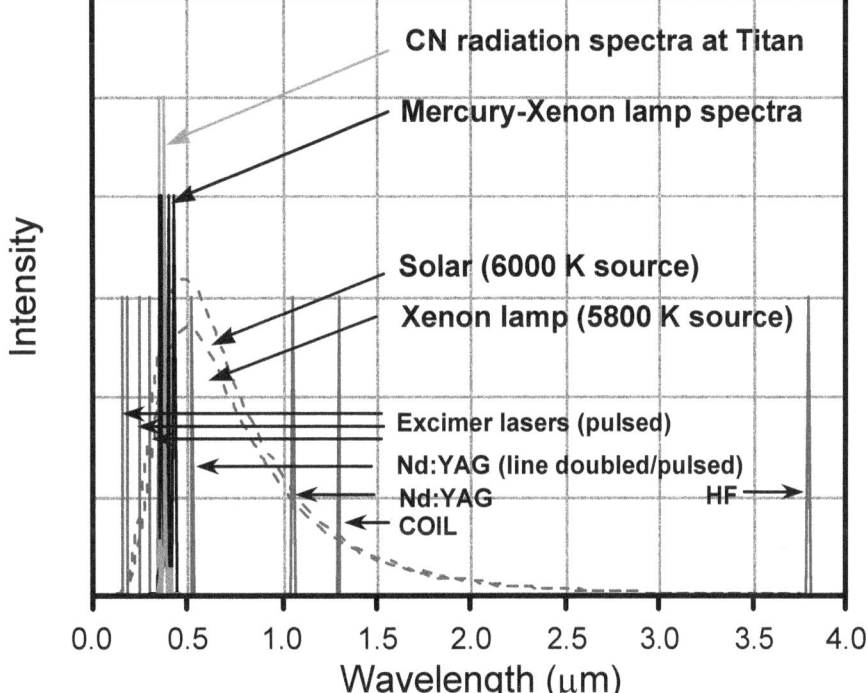

**Figure H-2.   Predicted Radiation for the Titan Probe and the Spectral Distribution for the
Test Facilities Considered**

The mercury-xenon lamp was delivered in February 2004.   Prior to delivery, NASA ARC fabricated and assembled hardware to enable a transverse subsonic flow of N_2 across the sample surface during the radiation testing. This UV test system was designed to operate at atmospheric pressure with N_2 gas blowing through a subsonic nozzle in a shear flow across the test model surface. A N_2 shear flow was incorporated for several reasons:

- To more accurately simulate the non-oxidizing Titan atmosphere.

- To prevent the decomposition products from blocking or absorbing the incident radiation beam.

- To prevent deposition on and subsequent contamination of the optical components used to focus the ultraviolet beam.

N_2 was supplied from a cryogenic tank, run through an evaporator and high-pressure lines to a subsonic nozzle or mini-wind tunnel, which was designed to provide flow rates of Mach 0.1 at the gas nozzle exit, at the lower edge of the test model. Convective cooling from the N_2 gas flow

	NASA Engineering and Safety Center Technical Assessment Report	Document #: RP-05-67	Version: 1.0
Title: **Independent Technical Assessment of Cassini/Huygens Probe Entry, Descent and Landing (EDL) at Titan**			Page #: 102 of 116

was anticipated in the experimental design, but it was not considered significant because it does not affect the radiation transport through the material. In this context, the N_2 gas convection merely reduces the model surface temperature.

A data acquisition system was also assembled to record the time-dependent thermocouple and pyrometer data to be acquired during test.

H.3 UV Test Procedure

Each sample of AQ60 was instrumented with in-depth thermocouples (T/C). Some samples had T/Cs at depths of 1, 2, 3, and 12 mm from the surface with a fifth thermocouple on the rear surface of the sample. Some samples only had T/Cs at 1 and 3 mm with a third thermocouple on the backwall and the rest only had backwall T/Cs.

The test procedure involved measuring the heat flux with a Gardon Gage calorimeter while the N_2 flow was active and then replacing the Gardon Gage with a material sample, establishing the N_2 flow, and finally irradiating the sample at the desired heat flux and duration.

The surface temperature was monitored with an optical pyrometer when possible. Based on the radiative heating estimates generated during the Titan aerocapture systems analysis study, heating rates between 50 and 150 W/cm2 were the desired levels. This range also encompassed the best estimate of radiative heating during direct entry of the Huygens probe.

H.4 UV Tests

H.4.1 UV Test at 50 W/cm^2

The first test of AQ60 was conducted in August 2004 at a nominal heat flux of 50 W/cm^2 for 150 seconds duration. The clearance was sufficient between the optics and the test area to allow a good view factor between the test model, the support structure, and UV protective shields, in order to take advantage of both surface pyrometry and to make a high-speed video of the test runs.

An infrared pyrometer, Model M190RH, was used to record the surface temperature during the test runs, which enabled comparisons between surface temperature with that measured by subsurface T/Cs at 1, 2 and 3 mm. These were the first measurements made of such precision for this class of low-density ablators. Steep temperature gradients across these small distances demonstrated that the material was performing very well.

The conditions were calibrated and verified using a Gardon Gage Calorimeter with a newly coated sensor element using SIC (DAP BBQ Black) to give a stable high-temperature 0.9

emittance coating, as confirmed by spectrophotometer measurements over the wavelength range of interest.

Figure H-3 shows the thermocouple data taken during the nominal heat flux test at 50 W/cm^2 which shows excellent behavior of AQ60 under these test conditions. The figure also shows good surface absorption in the char layer and no evidence of transparency or in-depth absorption. Figure H-4 provides pre- and post-test photos of the sample.

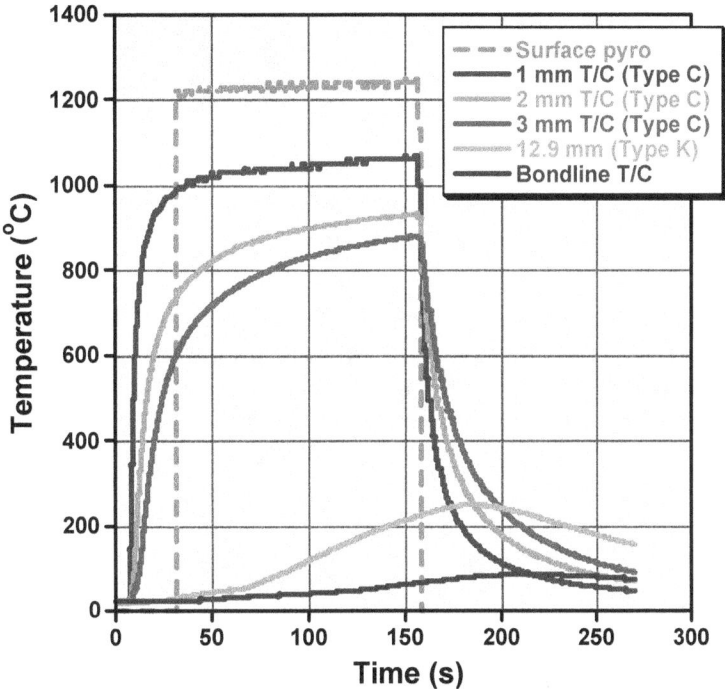

Figure H-3. Thermal Data from UV Test at 50 W/cm^2

Figure H-4. Photographs of AQ60 *Before* and *After* at Exposure at 50 W/cm² for 150 Seconds

When the data was transmitted to the ESA, Jean-Marc Bouilly (EADS) commented that the pyrometer data were lower than he would expect at that heat flux. Indeed, the surface temperatures measured in AQ60 during the nominal heat flux tests at 50 W/cm2, were lower than predicted from simple back-of-the-envelope calculations made assuming radiative equilibrium. That is, during the 50 W/ cm² test runs, although the energy flux was measured as 50 W/cm² by the Gardon Gage calorimeters used, AQ60 surface temperatures of 1200-1250°C implied an incident flux of 30 W/cm² using simple radiative equilibrium. This difference was attributed to convection cooling from the N_2 shear flow.

H.4.2   Convective Cooling Effects

The different geometry of the calorimeter and test model (small hemisphere vs. larger flat plate geometry) did not allow a one-to-one scaling comparison of the measured convective cooling. Simple analytical or computational models could not be applied given the lack of detailed property data for this non-NASA material, and the inherent complexity of combining ablation, radiation, and convection. Consequently, it was more efficient to run separate tests to put the issue to rest and to resolve a series of ARC/ESA discussions of the heat flux level long before the rapidly approaching date of deployment of the Huygens probe. Therefore, a series of tests was run to verify and quantify this convective cooling effect.

The series of flow/no-flow tests verified that the N_2 gas flow was entirely responsible for the lower surface temperatures measured in the test runs. Figure H-5 shows that when the N_2 gas was closed off and the model was surrounded by air at ambient pressure, driven only by natural convection and the chemical fume hood ventilation, the surface temperature rose to 1550°C. This was consistent with anticipated surface temperature at the intended 50 W/cm² flux as

NASA Engineering and Safety Center Technical Assessment Report	Document #: RP-05-67	Version: 1.0
Title: Independent Technical Assessment of Cassini/Huygens Probe Entry, Descent and Landing (EDL) at Titan		Page #: 105 of 116

measured by the calorimeter. When the N_2 gas was allowed to flow across, and convectively cool, the heated test material, the surface temperatures were reduced to 1250 °C, corresponding to an apparent net heat flux of 30 W/cm^2. This implied that convective cooling accounts for $\approx$ 20 W/cm^2.

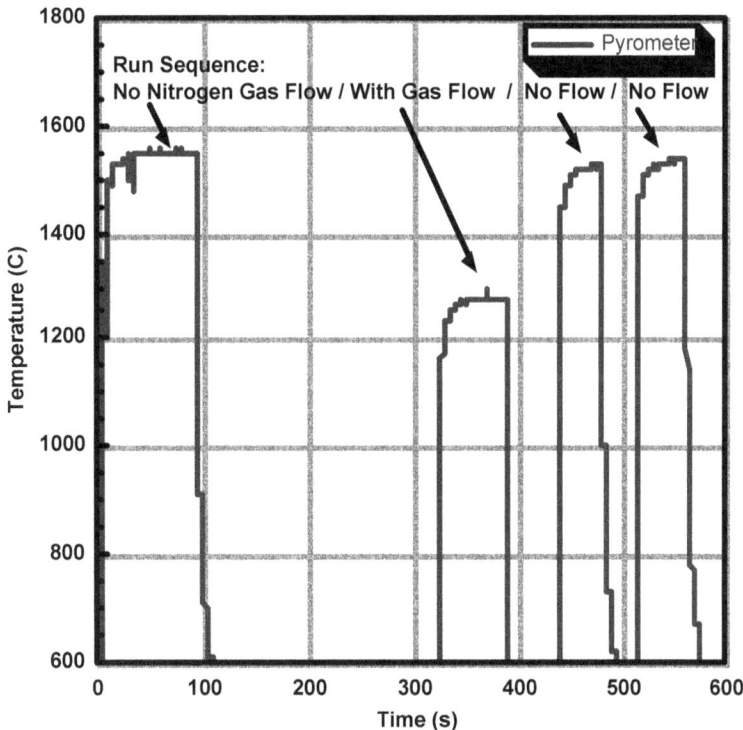

Figure H-5. Effect of N_2 Gas Flow across AQ60 char at 50 W/cm^2

H.4.3 UV Test at 150 W/cm2

A radiant heat flux of 150 W/cm^2 was considered as the maximum possible, including all safety margins of concern. Thirty seconds of duration was chosen to give a total heat load of 4500 J/cm^2. This was a severe over-test, but the material behaved very well.

Due to tight clearance and nonexistent direct line of sight to the test area, it was not possible to record the surface temperature using an infrared pyrometer, or to make a high-speed video of the test runs, as had been the procedure for the nominal heat flux runs. The clearance was minimal (0.75 cm between the optics and the test area) because the system had been reconfigured and additional optics had been introduced to get the required energy fluxes for the high heat flux

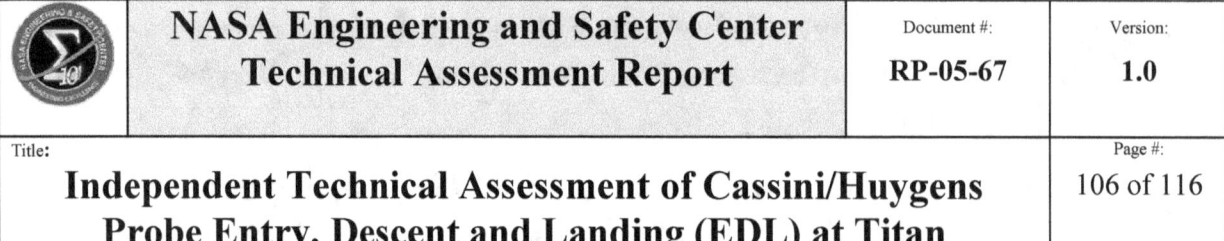

NASA Engineering and Safety Center Technical Assessment Report	Document #: RP-05-67	Version: 1.0
Title: **Independent Technical Assessment of Cassini/Huygens Probe Entry, Descent and Landing (EDL) at Titan**		Page #: 106 of 116

runs. This was not considered crucial because earlier tests had established the relationship between the surface temperatures and subsurface thermocouple data.

Figure H-6 shows the thermocouple data taken during the high heat flux test at 150 W/cm^2 which shows excellent behavior of AQ60 under these test conditions. The figure also shows good surface absorption in the char layer and no evidence of transparency or in-depth absorption. If there was in-depth absorption, it would be reflected by an instantaneous vertical rise of the 1st in-depth thermocouple to the temperature limit of the thermocouple.

Pre- and post-test photos are shown in Figure H-7.

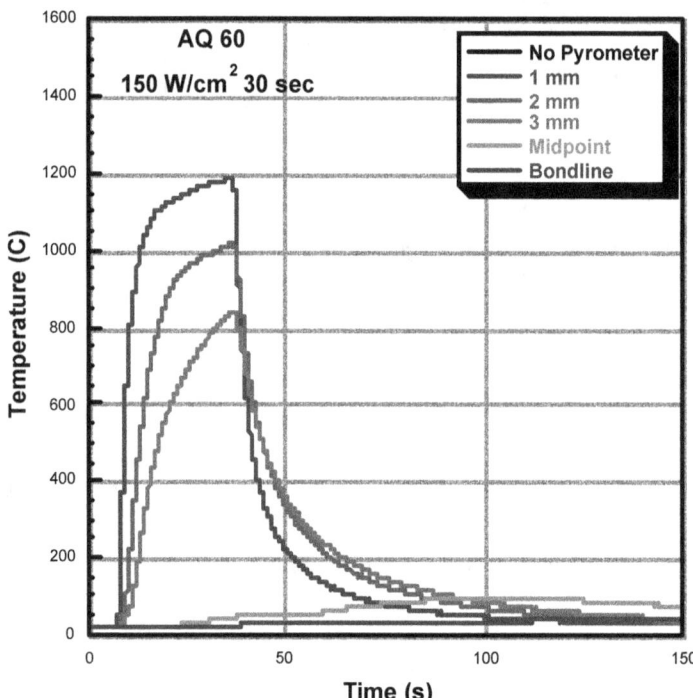

Figure H-6. Test Data Showing In-Depth Thermocouple Response at 150 W/cm^2

NASA Engineering and Safety Center Technical Assessment Report	Document #: RP-05-67	Version: 1.0
Title: **Independent Technical Assessment of Cassini/Huygens Probe Entry, Descent and Landing (EDL) at Titan**		Page #: 107 of 116

 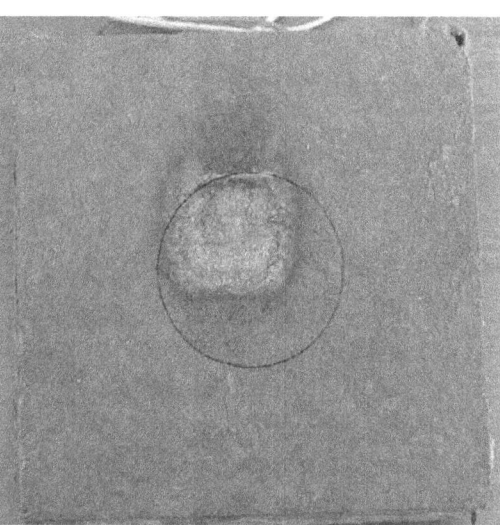

Figure H-7. Photos of AQ60 *Before* and *After* Test at 150 W/cm² for 30 Seconds

H.5   Arc Jet Test

Given the very good performance of AQ60 in the UV tests, it was decided to convert one of the UV test samples to a sample geometry appropriate for arc jet testing at relevant Huygens entry conditions. Consequently, one of the 75 mm square samples was machined to a circular geometry. The plug was instrumented with in-depth thermocouples at depths of 2.54, 5.08, 7.62 and 10.16 mm from the surface plus a T/C at the bondline between the rear of the sample and a 0.32 cm thick 2024 aluminum backplate. The AQ60 sample was 18.36 mm thick. The machined 76.7 mm diameter circular sample was embedded in a 101.6 mm diameter TUFI-coated high-density tile that is a sample size compatible with the level of heating desired in the arc test.

This test was conducted in the NASA ARC AHF arc jet facility in mid-December 2004. The test was conducted in a N_2 flow at a cold-wall stagnation point heat flux of 80 W/cm² for an exposure time of 57 seconds. Stagnation pressure was 2.38 kPa. Surface temperature was monitored with two pyrometers. After extraction from the flow, the sample was allowed to cool in a very low pressure environment while the bondline continued to rise in temperature due to heat soak.

No surface recession was evident in this test. Again, very high quality thermocouple data were acquired as shown in Figure H-8. Pre- and post-test photos are shown in Figure H-9.

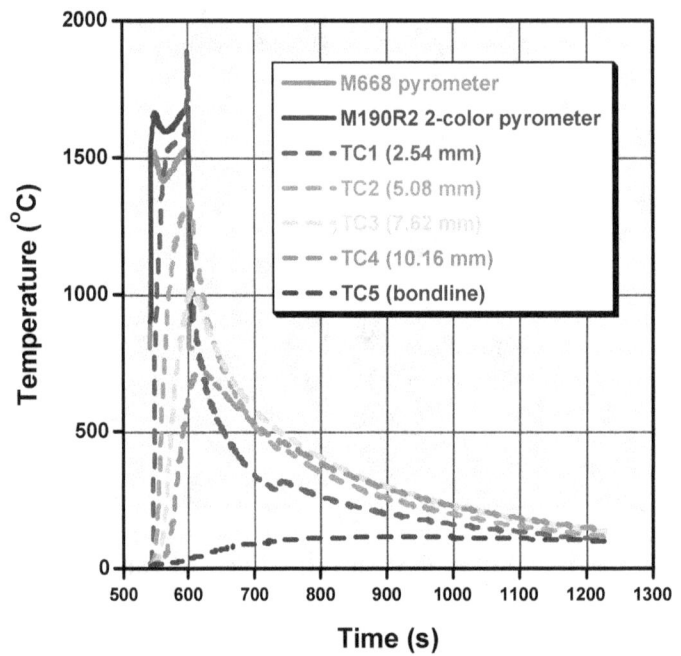

Figure H-8. Pyrometer and Thermocouple Data from Arc Jet Test of AQ60

Figure H-9. Photos of AQ60 *Before* and *After* Arc Jet Exposure at 80 W/cm^2

NASA Engineering and Safety Center Technical Assessment Report	Document #: RP-05-67	Version: 1.0
Title: **Independent Technical Assessment of Cassini/Huygens Probe Entry, Descent and Landing (EDL) at Titan**		Page #: 109 of 116

Not too much credence should be given to the magnitude of the maximum bondline temperature since the aluminum backplate installed for this test has a significantly higher thermal capacitance than a typical flight structure. There should not be too much credence given to the maximum bondline temperature in the arc jet test as it is not representative of flight. This is because the backup structure in the arc jet test has much higher thermal capacitance in comparison to the flight structure. Consequently, the maximum temperature at the bondline in the arc jet test is lower than would be anticipated in flight. This test demonstrated that the AQ60 performs reliably at this condition. More importantly, the thermocouple data is critical to development of a high-fidelity thermal/ablation model for this material. Unfortunately, schedule and budget constraints did not permit that activity to be completed.

H.6   Analytical Modeling

To support the NESC risk assessment, NASA ARC reviewed available information on the Huygens forebody TPS material, AQ60, to evaluate whether the models employed for TPS design were conservative.

H.6.1 Recession Modeling

AQ60 is a felt made of short silica fibers reinforced by impregnation of phenolic resin (30% by mass). The density of the virgin material is ≈ 0.280 g/cm^3 and the density of the char (after pyrolysis of the phenolic resin) is ≈ 0.240 g/cm^3.

During development of the Huygens probe, arc jet tests were conducted in facilities at both IRS Stuttgart and Aerospatiale Simoun. The tests at IRS Stuttgart were stagnation tests over a broad range of heat fluxes (≈ 60 to 250 W/cm^2) in a simulated Titan gas mixture of 77% N_2, 20% Ar and 3% CH_4 (by volume). Two tests were conducted in a pure N_2 flow. EADS developed an empirical curve-fit of the recession data as shown in Figure H-10.

NASA Engineering and Safety Center
Technical Assessment Report

Document #:	Version:
RP-05-67	**1.0**

Title:

Independent Technical Assessment of Cassini/Huygens Probe Entry, Descent and Landing (EDL) at Titan

Page #:

110 of 116

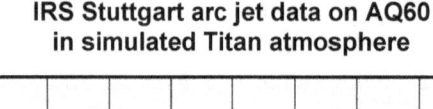

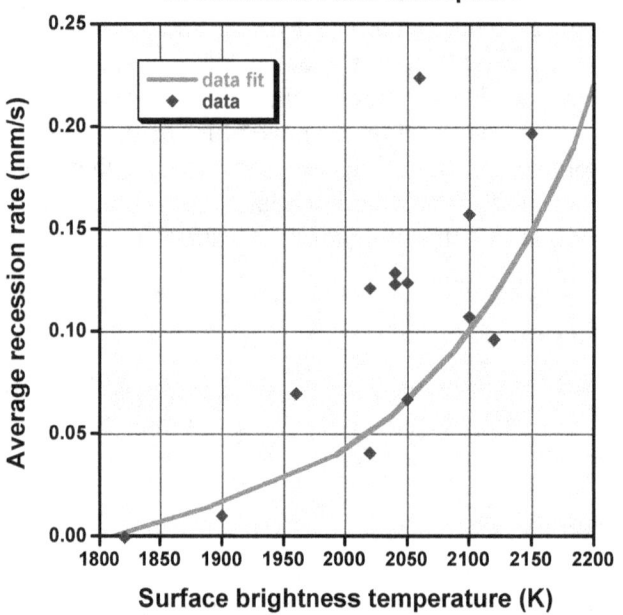

Figure H-10. Empirical Correlation of AQ60 Arc Jet Recession Data

All of the IRS Stuttgart arc jet tests were conducted at a stagnation pressure of ≈ 16 mbar (≈ 0.021 atm). The stagnation pressure anticipated in flight is predicted to be 2 to 4 times greater than that experienced in the IRS Stuttgart arc jet tests.

.

Ablative material performance is dependent on pressure. The ESA empirical correlation of recession data does not account for the effects of pressure on the ablation mechanism and, as such, introduces risk when extrapolated to conditions beyond the range of the data it is based on. To minimize this risk, we attempted to model the recession data from the IRS Stuttgart arc jet tests with a thermochemical ablation model since such a theoretical model would demonstrate the effects of pressure on recession rates.

We explored a range of viable possibilities. Since the phenolic pyrolyzes leaves a carbonaceous char at the surface, it is possible that carbon ablation is the controlling mechanism. However, the material is primarily silica fibers so glass melt, flow, and/or vaporization are more likely mechanisms.

NASA Engineering and Safety Center
Technical Assessment Report

Document #: RP-05-67

Version: 1.0

Title:
Independent Technical Assessment of Cassini/Huygens Probe Entry, Descent and Landing (EDL) at Titan

Page #: 111 of 116

Thermochemical equilibrium solutions are sensitive to the condensed phase chemical species at the heated surface. A glass-phenolic composite exposed to a N_2-Ar-methane flow field can form several surface species including C^*, Si^*, SiO_2^*, SiC^*, and $Si_3N_4^*$, where the asterisk indicates a condensed (solid or liquid) phase. Thermochemical equilibrium surface ablation solutions were generated with the ACE [ref. 1] code for a range of assumptions on possible surface species. Candidate surface species can be easily eliminated from consideration by removing the thermochemical data [ref. 2] for that species from the input file.

Surface recession is expressed in terms of a parameter called B'_c which is a non-dimensional char mass loss rate. The influence of pyrolysis gas injection into the boundary layer is shown in terms of B'_g, which is a non-dimensional pyrolysis gas mass loss rate. The predicted variation of B'_c with surface temperature had a very different character dependent on which candidate surface species were considered. In some cases, the $B'_c - T_W$ solutions appeared reasonable, but the surface temperatures were too low, i.e., much lower than the pyrometer brightness temperatures measured in the arc jet tests at IRS Stuttgart. The most reasonable solutions were obtained when candidate surface species were limited to Si^*, SiO_2^* and $Si_3N_4^*$. There is some difference in the solutions if $Si_3N_4^*$ is also eliminated from consideration, but at the lower end of the temperature range.

We concluded that the solutions where Si^* and SiO_2^* were the only candidate surface species that provided the best representation of the data. The data had to be recast in terms of B'_c, but this was straightforward. Furthermore, it was assumed that the pyrolysis gases were also in thermochemical equilibrium.

Figure H-11 illustrates a comparison of the thermochemical equilibrium solutions for AQ60 in the simulated Titan atmosphere in the IRS Stuttgart arc jet tests. The data are also shown with ±5% error bars on surface temperature. In this figure, the brightness temperatures have been corrected for an assumed surface emissivity of 0.9.

Figure H-11. Comparison of Thermochemical Ablation Model with IRS Arc Jet Data in Simulated Titan Atmosphere

As seen, the thermochemical equilibrium solutions exhibit a trend with surface temperature that is consistent with the data and, in fact, are as good a representation as the EADS empirical correlation. Solutions are only shown for a limited range of B'_g since we estimate that the range of B'_g values is limited and confined to relatively small values. This model is based on the premise that the controlling ablation mechanism is the melting and vaporization of the glass fibers in AQ60, but not the melt runoff. The post-test pictures of the samples in the IRS Stuttgart arc jet show no evidence of molten glass flow.

It is also useful to see if this ablation model also predicts the two tests conducted in the N_2-only flow. As seen in Figure H-12, the correlation is very good. It is unfortunate that these two tests were conducted at the same condition since it would be more valuable to see if the ablation of AQ60 in N_2 had the same trend as the model. This would require tests over a broader range of conditions.

| NASA Engineering and Safety Center | Document #: | Version: |
| Technical Assessment Report | RP-05-67 | 1.0 |

| Title: Independent Technical Assessment of Cassini/Huygens Probe Entry, Descent and Landing (EDL) at Titan | Page #: 113 of 116 |

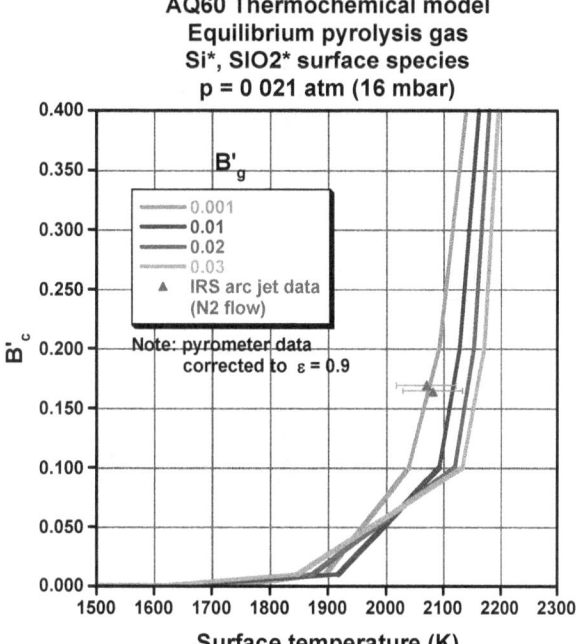

Figure H-12. Comparison of Thermochemical Ablation Model with IRS Arc Jet Data in N_2

The thermochemical model attributes the ablation of AQ60 to the vaporization of silica (the predominant component in AQ60), but not melt flow. The advantage of a thermochemical model is that it can be extrapolated to other conditions with some confidence whereas a model based on an empirical correlation has validity only within the range of the data it is based on.

This is demonstrated in Figure H-13 where non-dimensional char mass loss rates were calculated for a different assumed atmospheric composition (95% N_2, 5% CH_4 by volume) and at two pressures (0.021 atm and 0.084 atm). As seen, higher pressures will force ablation to occur at higher surface temperatures. This means that for equivalent heating rates, surface recession at higher pressures will be less than recession at lower pressures. This does not necessarily mean that TPS performance at higher pressure would be improved since less recession means less energy absorbed in ablation (e.g., the heat of vaporization of glass is $\approx$ 12.5 kJ/g). An assessment of this effect requires detailed analysis of TPS performance for a given heat pulse.

NASA Engineering and Safety Center Technical Assessment Report	Document #: RP-05-67	Version: 1.0
Title: Independent Technical Assessment of Cassini/Huygens Probe Entry, Descent and Landing (EDL) at Titan		Page #: 114 of 116

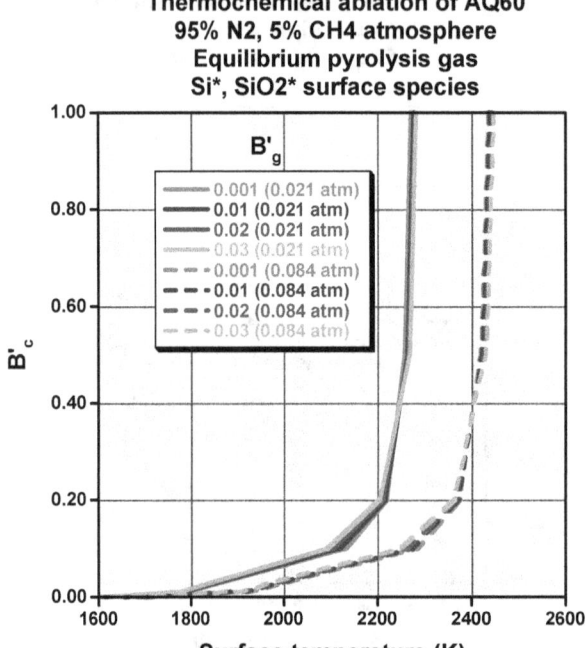

Figure H-13. Comparison of Non-Dimensional Mass Loss Rates at Two Pressures

H.6.2 Thermal Modeling

The EADS thermal model employed for TPS design, as described in a variety of EADS documents, is an empirically-based model that was assembled from test data acquired during development of the Huygens probe. This includes both arc jet tests, which did not include in-depth thermocouples, and tests with infrared lamps, which did include in-depth thermocouples. By NASA standards, this model would not be considered a high-fidelity thermal ablation model.

It is difficult to evaluate a model without specific knowledge of how the model developer uses it. At ARC, we took the EADS model and compared predictions with the thermocouple data from the UV tests with reasonably good correlation. In contrast, similar comparison with the thermocouple data from the arc jet test was not encouraging as the model predictions were below the data. However, these studies were very limited in scope so it would be unfair to draw any conclusions.

References

1. Anon., "User's Manual, Aerotherm Chemical Equilibrium Computer Program (ACE81)," Acurex Corporation, Aerotherm Division, Mountain View, California, August 1981.

2. "JANAF Thermochemical Tables," Third Edition, *J. of Physical and Chemical Reference Data,* Vol. 14, 1985.

Approval and Document Revision History

Approved:	Original signature on file	6-5-05
	NESC Director	Date

Version	Description of Revision	Office of Primary Responsibility	Effective Date
1.0	Initial Release	Flight Science Discipline Expert Office	6-5-05